Advanced Ceramics and Composites

Volume 7

Series Editor

Longbiao Li, College of Civil Aviation, Nanjing University of Aeronautics and Astronautics, Nanjing, Jiangsu, China

The book series "Advanced Ceramics and Composites" publishes insights and latest research results on advanced ceramics and composites, as well as the applications of these materials. The intent is to cover all the technical contents, applications, and multidisciplinary aspects of Advanced Ceramics and Composites.

The objective of the book series is to publish monographs, reference works, selected contributions from specialized conferences, and textbooks with high quality in the field advanced ceramics and composite materials. The series provides valuable references to a wide audience in research community in materials science, research and development personnel of ceramic and composite materials, industry practitioners and anyone else who are looking to expand their knowledge of ceramics and composites.

Xiaochong Liu · Longbiao Li

Design, Fabrication and Testing of Aeroengine Ceramic-Matrix Composite Components

Springer

Xiaochong Liu
School of Materials Science and Engineering
Northwestern Polytechnical University
Xi'an, Shaanxi, China

Longbiao Li
College of Civil Aviation
Nanjing University of Aeronautics and Astronautics
Nanjing, Jiangsu, China

ISSN 2662-9305 ISSN 2662-9313 (electronic)
Advanced Ceramics and Composites
ISBN 978-981-97-6108-1 ISBN 978-981-97-6109-8 (eBook)
https://doi.org/10.1007/978-981-97-6109-8

This Springer imprint is published by the registered company Springer Nature Singapore Pte Ltd.
The registered company address is: 152 Beach Road, #21-01/04 Gateway East, Singapore 189721, Singapore

Preface

Ceramic-matrix composites (CMCs) can withstand higher temperatures, reduce cooling airflow, and improve turbine efficiency compared to high-temperature alloys. CMCs have been applied in hot-section components of aeroengines, e.g., combustor liners, turbine guide vanes, turbine blades, turbine blisk, exhaust mixer, nozzles, etc. This book focuses on the design, fabrication and testing of aeroengine CMC components, as following:

- The mechanical performance and fabrication methods, e.g., chemical vapor infiltration (CVI), polymer and infiltration pyrolysis (PIP), and melt infiltration (MI), are introduced. The requirements of aeroengine hot-section components are given and related with CMCs. The applications of CMC components in stator and rotation hot-section components of aeroengines are introduced.
- The high-temperature mechanical properties of chemical vapor infiltration (CVI) processed SiC/SiC composites with different fiber preforms, i.e., three-dimensional four-directional (3D4D) braided, two-dimensional plain-woven (2DPW) and stain woven (2DSW) lamination, are investigated at elevated temperature of 1400 °C. The high-temperature mechanical properties of SiC/SiC composites, e.g., tensile strength, compressional strength, interlaminar shear strength, in-plane shear strength, flexural strength, cyclic fatigue lifetime, and creep life are obtained from experimental data. The high-temperature mechanical properties can support the engineers to better design CMCs hot-section components.
- The design, fabrication and testing of CMC high-pressure single/twin turbine guide vanes were investigated. The service performance analysis of the turbine guide vane was at first analyzed. The SiC/SiC high-pressure turbine single/twin guide vanes were fabricated using the chemical vapor infiltration (CVI) method. Thermal shock experiments of SiC/SiC high-pressure turbine guide vanes were carried out on the third-generation turbine service environment simulation device. For single turbine guide vane, the thermal shock tests between 350 and 1350 °C in gas environment were conducted; and for twin turbine guide vane, cyclic thermal shock tests at different target temperatures (i.e., 1400, 1450, and 1480 °C) in gas

environment were conducted to investigate the damage mechanisms and failure modes. During thermal shock tests, the surface temperature field of the SiC/SiC guide vane was analyzed. The damage characteristics of the single/twin turbine guide vane after experiencing different thermal shocks were discussed.

- The design, fabrication and testing of CMC combustor liner were investigated. The service performance analysis of the combustor liner was at first analyzed. The SiC/SiC combustor liner with the dimension of 296.48 mm (maximum diameter of outer liner) × 149.85 mm (minimum diameter of inner liner) × 100.46 mm (height), a wall thickness of 3–10 mm (including splicing structure) was fabricated using the CVI process. The surface of the SiC/SiC combustor liner was coated with EBC (Environmental Barrier Coatings, EBC) to improve the reliability and durability. Pulsed laser micromachining technology was used to realize the precise processing of SiC/SiC combustor liner micro-holes, and ultrasound-assisted machining technology was used to meet the demand for high-precision profile processing for various air-film holes with diameters larger than 0.5 mm, and the roughness can reach Ra = 0.3~0.5. After the inner and outer liners and other components have been densified and processed, the SiC rivets are used for on-line riveting, which achieves a connection result equivalent to "metal riveting" after CVI-SiC deposition, improves the ability of components to maintain the profile, and prevents delamination of combustor liner during the service life. On October 14, 2022, the ground ignition and performance test of the combustor liner was conducted. The average gas temperature was 1426 °C, with localized hot spots reaching 1734 °C. Localized EBC coating peeling was found after the ground testing.
- The design, fabrication and testing of CMC turbine blisk were investigated. The service performance analysis of the turbine blisk was at first analyzed. The spider web structure (SWS) fiber preform was adopted as the reinforcement elements of the turbine blisk. The BN interphase and SiC matrix were deposited on the SWS fiber preform, and during the fabrication, the on-line machining were conducted to control the shape and dimension of the blisk. The environmental barrier coatings (EBCs) were prepared on the surface of SiC/SiC turbine blisk using the atmospheric plasma spraying (APS) method. The XCT scan was conducted to characterize the internal defects in SiC/SiC turbine blisk. The mechanical performance evaluation, over rotation test and engine bench test were performed, respectively. Finally, the SiC/SiC turbine blisk was installed on an aircraft and completed the first flight testing, which verified the feasibility of the SiC/SiC turbine rotor for gas turbine engine application and accumulated valuable experience for the subsequent development of CMC hot-section components.

We hope this book can help the engineering designers to understand and master the design, fabrication and testing of CMC hot-section components in aeroengine.

Xi'an, China
Nanjing, China
March 2024

Xiaochong Liu
Longbiao Li

Contents

Chapter 1
Introduction

Abstract In this chapter, the mechanical performance and fabrication methods, e.g., chemical vapor infiltration (CVI), polymer and infiltration pyrolysis (PIP), and melt infiltration (MI), are introduced. The requirements of aeroengine hot-section components are given and related with CMCs. The applications of CMC components in stator and rotation hot-section components of aeroengines are introduced.

Keywords SiC/SiC · Mechanical performance · Fabrication method · Application

1.1 Mechanical Performance of Ceramic-Matrix Composite

Ceramic-matrix composites (CMCs) not only retain the excellent properties of ceramic materials, such as high temperature resistance, low density, high specific strength, high specific modulus, oxidation and ablation resistance, but also overcome the fatal weaknesses of ceramic materials, such as high brittleness and poor reliability (Li 2021). According to the different toughening methods, CMCs are classified into four categories: particle, whisker, laminated and continuous fiber toughened, with increasing strength and fracture toughness in that order (Li 2024). Depending on the composition of CMCs, continuous fiber-reinforced CMCs are classified into three categories: glass-matrix, oxide-matrix, and non-oxide-matrix, with working temperatures increasing sequentially. Oxide ceramic-matrix composites (Oxide-CMCs) are more stable in oxidizing environments, but their mechanical properties, especially resistance to high-temperature creep, are poor, so non-oxide ceramic matrix composites (Non-oxide CMCs) have gained wider attention than the former (Li 2023). Among the non-oxide CMCs, fiber-reinforced silicon carbide ceramic-matrix composites (SiC-CMCs) have received the most attention. SiC-CMCs have outstanding advantages such as metal-like fracture behavior, insensitivity to cracks, no catastrophic damage, etc. They are able to meet the requirements of long life below 1650 °C, limited life below 2000 °C, and instantaneous life below 2800 °C (Li 2022). SiC-CMCs mainly include carbon fiber reinforced silicon carbide

X. Liu and L. Li, *Design, Fabrication and Testing of Aeroengine Ceramic-Matrix Composite Components*, Advanced Ceramics and Composites 7,
https://doi.org/10.1007/978-981-97-6109-8_1

composites (C/SiC) and silicon carbide fiber reinforced silicon carbide composites (SiC/SiC).

The mechanical properties of ceramic matrix composites (CMCs) reflect the damage resistance of composites in different stress states, and are important parameters for the safe design of CMCs hot-section components, which are directly related to the reliability, safety and durability of the components in service (Liu et al. 2023b). The basic mechanical properties of CMCs mainly include in-plane tensile properties, interlaminar shear strength, in-plane shear strength, compression strength, flexural strength, fatigue and creep properties, etc. Table 1.1 gives the test standards of mechanical properties of CMCs. Table 1.2 lists the main mechanical properties of 2D CVI SiC/SiC composites at room temperature.

Figure 1.1 shows experimental tensile stress–strain curve and AE energy versus loading time curves of 2D CVI SiC/SiC composite (Guo et al. 2021). Under tensile loading, the composite exhibits obviously nonlinear behavior due to multiple damage mechanisms of matrix cracking, interface debonding, and fibers broken.

Table 1.1 ASTM standards of testing mechanical properties of ceramic-matrix composites

Mechanical property	Testing temperature	ASTM standard
Tensile	Ambient temperature	ASTM C1275-18 standard test method for monotonic tensile behavior of continuous fiber-reinforced advanced ceramics with solid rectangular cross-section test specimens at ambient temperature
	Elevated temperature	ASTM C1359-18e1 standard test method for monotonic tensile strength testing of continuous fiber-reinforced advanced ceramics with solid rectangular cross section test specimens at elevated temperatures
Interlaminar shear strength	Ambient temperature	ASTM C1292-22 standard test method for shear strength of continuous fiber-reinforced advanced ceramics at ambient temperatures
	Elevated temperature	ASTM C1292-22 standard test method for shear strength of continuous fiber-reinforced advanced ceramics at ambient temperatures
In-plane shear strength	Ambient temperature	ASTM C1292-22 standard test method for shear strength of continuous fiber-reinforced advanced ceramics at ambient temperatures
Compressive strength	Ambient temperature	ASTM C1358-18 standard test method for monotonic compressive strength testing of continuous fiber-reinforced advanced ceramics with solid rectangular cross section test specimens at ambient temperatures
Flexural strength	Ambient and elevated temperatures	ASTM C1341-13 (2018) standard test method for flexural properties of continuous fiber-reinforced advanced ceramic composites

Table 1.2 Mechanical properties of 2D CVI SiC/SiC composite at room temperature

Mechanical properties	2D CVI SiC/SiC
Tensile strength (MPa)	243.5
Tensile modulus (GPa)	281.6
Compressive strength (MPa)	672.3
Compressive modulus (GPa)	287
Flexural strength (MPa)	485
Flexural modulus (GPa)	109.2
In-plane shear strength (MPa)	212.5
In-plane shear modulus (GPa)	120
Interlaminar shear strength (MPa)	53
Fracture toughness (MPa $m^{1/2}$)	26.5

The composite elastic modulus was approximately E_c = 132 GPa. Based on the monitoring of acoustic emission, the tensile curve can be divided into three stages, including:

(1) Stage I, which starts from initial loading to applied stress $\sigma_{mc} = 50$ MPa. At this stage, the AE count and AE energy are in the middle level, the duration and amplitude are basically stable after a period of rising range, and the frequency range of AE is wide, which indicates that the specimen gradually produces damage in the initial stage of loading, but the energy of single damage is small, mainly due to initial micro cracking and partial interface debonding.
(2) Stage II, which lies in the stress range from $\sigma_{mc} = 50$ MPa to $\sigma_{sat} = 215$ MPa. At this stage, the AE count is in the middle level, and the AE energy is relatively high. The frequency of AE signals lies mainly between 0–70 kHz and 130–220 kHz, which indicates that the matrix cracks continue to propagate and a small amount of fiber fracture occurs.
(3) Stage III, which lines in the stress range from $\sigma_{sat} = 215$ MPa to $\sigma_{uts} = 266.78$ MPa. At this stage, the AE signal count and AE signal energy are relatively low and accompanied with a small amount of medium level energy release. Except for the high energy release in a short time, the frequency range of AE is wide, and the amplitude is low. This indicates that when some longitudinal fibers fracture, the material is in a stable state, except for a small amount of matrix crack propagation and fiber fracture, the sample has little damage until tensile fracture.

In Chap. 2, the high-temperature mechanical properties of chemical vapor infiltration (CVI) processed SiC/SiC composites with different fiber preforms, i.e., three-dimensional four-directional (3D4D) braided, two-dimensional plain-woven (2DPW) and stain woven (2DSW) lamination, are investigated at elevated temperature of 1400 °C.

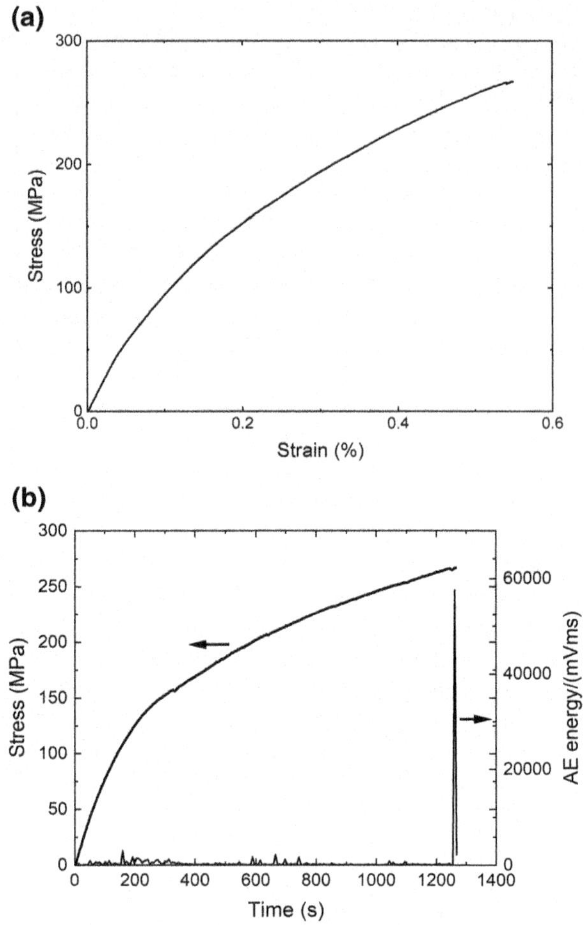

Fig. 1.1 **a** Experimental and predicted tensile stress–strain and interface debonding ratio curves; and **b** the stress and acoustic energy versus loading time curves of 2D CVI SiC/SiC composite

1.2 Fabrication Methods of Ceramic-Matrix Composites

After years of development, the CMC fabrication process has matured, and the current main fabrication processes include the following three:

- Chemical Vapor Infiltration (CVI)
- Precursor Infiltration Pyrolysis (PIP)
- Melt Infiltration (MI)

1.2.1 Chemical Vapor Infiltration (CVI)

The CVI method was derived from CVD, a well-established technique for applying thin ceramic coatings. Originally developed for the manufacture of C/C materials, the CVI process could be adapted to the deposition of SiC matrix using a defined mixture of methyltrichlorosilane (MTS) and H_2 as process gas.

$$CH_3SiCL_3(g) + H_2(g) \rightarrow SiC(s) + 3HCl(g) + H_2(g)$$

The CVI technique can be used to manufacture simple plates, large and complex structures. The CVI fabrication process consists of chemical vapor infiltration of a fiber coating and a ceramic matrix into a two-dimensional (e.g., laminated fabric) or three-dimensional fiber architecture. During the CVI process, the gaseous precursors are pumped into a reactor containing a heated preform, and infiltrate the preform and react at the surface of the fibers, initially forming an interphase on the fiber, followed by the deposition of the matrix material. The porosity of CMCs after densifying the fiber preform can approach 10–20%. The deposition of the matrix is conducted at a low enough temperature in the range of 900–1100 °C to allow a uniform deposition of the matrix.

The CVI process can be divided into three main process steps, as follows,

(1) Manufacture and stabilizing of a fiber preform.
(2) Deposition of fiber coating.
(3) Deposition of the SiC matrix.

The main advantages of the CVI process are the relatively low process temperatures, no damage to the fibers and the high purity of the obtained SiC matrix, leading to the high performance of the CMC materials. Complex-shaped lightweight structures can be manufactured by in situ joining of simple substructures, and bonded together with SiC at an intermediate step of matrix densification. The main drawback of the CVI methods are the high investment cost for the facilities and the demanding process control and long process times and material inhomogeneities. Other disadvantages of CVI are the limited fiber content and the remaining open porosity.

1.2.2 Polymer and Infiltration Pyrolysis (PIP)

The characteristic of the PIP process is the matrix fabricated through thermal decomposition of the preceramic polymer in multiple densification cycles. The fibers or fiber preforms are infiltrated by the liquid precursor in the first step. After curing of the polymer, the resulting thermoset matrix is converted into a ceramic matrix via heat treatment in a non-oxidizing atmosphere. SiC matrix are usually derived from polycarbosilanes with high yields of SiC. An amorphous SiC matrix is formed during pyrolysis at a relatively low temperature of about 850–1050 °C. To increase matrix density and temperature stability, the matrix can be transformed to crystalline β-SiC by subsequent heat treatment at temperatures of 1500–1600 °C in inert gas. The PIP technique is one of the most advanced methods for the manufacture of large and complex-shaped CMCs structures in the aerospace industry.

The PIP process can be subdivided into four basic process steps, as follows,

(1) Deposition of fiber coating (interphase).
(2) Manufacture of a CFRP preform.
(3) Pyrolysis of the CFRP preform.
(4) Densification via repeated polymer infiltration and pyrolysis.

Similar to the CVI process, the fibers are not influenced or damaged during the deposition of the SiC matrix. Easily manufactured substructures of simple geometries are manufactured separately in the first process steps and joined together in the course of the densification process. However, the PIP process are high costs for the preceramic precursors and the long manufacturing times of several weeks up to months, especially for large, complex-shaped structures, mainly caused by the multiple infiltration and pyrolysis steps. The interlaminar shear strength is usually low due to the high residual porosity, especially for the two-dimensional fiber reinforced CMCs.

1.2.3 *Melt Infiltration (MI)*

The characteristics of the MI method are the infiltration of a porous C/C preform with molten Si and the fabrication of the SiC matrix by an exothermic chemical reaction of the liquid silicon with the solid carbon. The MI technique is used for the manufacture of flat plates and complex shaped, and large structures.

The GE (General Electric Co.) melt infiltrated, prepreg SiC/SiC CMC has evolved from material process development through panel making and testing to component demonstration through shroud fabrication and engine testing, and finally to shroud production for a certified aeroengine. For the process of the GE prepreg SiC/SiC, the SiC fiber tows are coated with boron nitride (BN)-based fiber interphase and a Si_3N_4 over-coating using a proprietary CVD process. Unidirectional fiber tapes are made from coated tow using a drum winding process. The next step is binder burnout or pyrolysis, which converts the polymer into carbon. Densification of the pyrolyzed preform is through Si melt infiltration. A typical microstructure consists of a SiC/Si matrix and nominally 20–25% fibers.

High pressure turbine shrouds manufactured from the prepreg MI SiC/SiC have been in production since 2016 for the certified CFM LEAP engines used in the Airbus A320neo. The GE9X engine, which will power the new Boeing 777x, will have 42 nozzles and the combustor liners manufactured from the prepreg MI SiC/SiC.

1.3 Requirements of Aeroengine Hot-Section Components

Since the early 1940s, large thrust-to-weight ratio, high efficiency and long life have been the eternal pursuit of aeroengine research, and improve the turbine inlet temperature can directly enhance the comprehensive performance of aeroengine, therefore, in the past eight decades, the aeroengine turbine inlet temperature has increased dramatically. In the future, the turbine inlet temperature will reach more than 2000 K for thrust-to-weight ratio 12–15 aeroengine, and up to 2200–2450 K for thrust-to-weight ratio 15–20 aeroengine turbine, far beyond the temperature limit of high-temperature alloy materials. CMCs possess excellent high-temperature mechanical properties, low density and other advantages, which are regarded as the next generation thermal structure materials for aeroengines.

Aeroengine hot-section components operate in extreme thermomechanical environments with high temperatures, high pressures, severe gas thermal corrosion, and high aerodynamic and thermal fatigue loads. In order to meet the requirements of advanced aeroengines, CMC components should have the following properties:

(1) Excellent oxidation and corrosion resistance at high temperatures.
(2) Excellent creep rupture strength, thermal fatigue properties, vibration fatigue properties at high temperatures.

CMC components are being developed to enable significant increase in aeroengine performance and safety, and to reduce costs.

1.4 Application of CMCs Components in Aeroengines

CMC components provide opportunities for pursing "Revolutionary Propulsion Concepts" enabling new, higher efficiency systems that can operate at higher temperatures with increased safety, as follows,

(1) CMC components can increase the safety margin due to higher temperature capability and higher damping capacity, while minimizing system complexity (e.g., elimination of need for cooling, fewer parts, etc.) and component and system weight.
(2) Low density of CMCs can allow increased thrust to weight and minimizes effects on stability when material is lost from rotating components.
(3) CMC components can decrease costs via higher temperature capability, low part count (example–integrally blisk), and increased component life.

1.4.1 CMC Stator Components

At present, the SiC/SiC aeroengine components that have been fabricated and tested include combustor liners, turbine guide vanes, turbine blades, turbine blisk, exhaust mixer, nozzles, and other hot-section components.

UTRC (United Technologies Research Center) and P&W Canada validated the SiC/SiC combustor in a PW 200 series combustor (Bhatia et al. 2010), as shown in Fig. 1.2. The fill annular CMC combustor rig was engine tested for 250 cycles between idle (40,000 rpm) and full power (57,000 rpm) and severely tested the response of the CMC/metal interfaces to accelerated thermal cycling. The test was stopped after 250 cycles with no damage observed. The use of CMC combustor in PW200 series demonstrated a 40–50% improvement in pattern factor, a 30% reduction in NO_x at full power and a 20% reduction in CO at idle condition compared to the baseline metal combustor.

The SiC/SiC combustor liner fabricated by the Solar Turbine Company successfully passed the 10,000 h durability test, and the test results show that: after using the SiC/SiC liner, the CO emission of the engine is significantly reduced, and as of the end of 2001, the SiC/SiC combustor liner has been subjected to a cumulative total of 50,000 h of durability tests, and five field tests have also been completed.

On May19, 2015, LEAP-1A civil turbofan engine equipped with CMCs high-pressure turbine shroud successfully completed its maiden flight on the Airbus A320neo aircraft, indicating a new breakthrough in the application of CMCs in the application of aeroengine hot-section components. This is the first time that CMCs have been used in engine core components, and the weight of the components has been reduced by hundreds of kilograms compared with traditional metal materials. 20,000 h of component and aeroengine tests have been completed on CMCs high-pressure turbine shroud, which have shown that the application of the CMCs components greatly reduces the amount of cold air from the compressor, improves the engine's thrust, and reduces the fuel consumption by more than 1.5%.

The GE9X engine to be installed on the Boeing 777X-9 utilizes SiC/SiC combustor inner and outer liners, first and second stage turbine guide vanes, and a high-pressure turbine shroud, and has been described as the commercial aeroengine that applies the most CMCs.

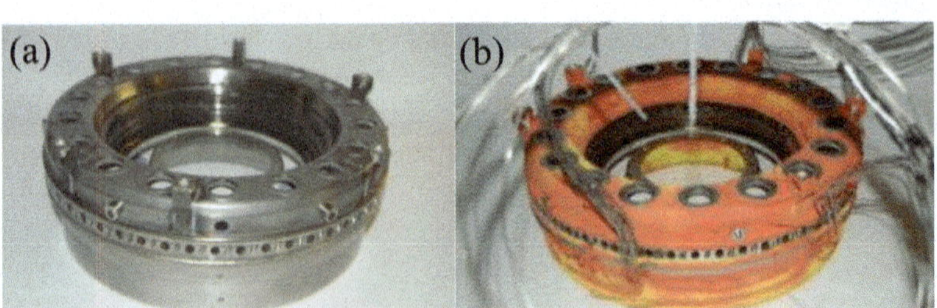

Fig. 1.2 MI SiC/SiC combustor before and after rig test

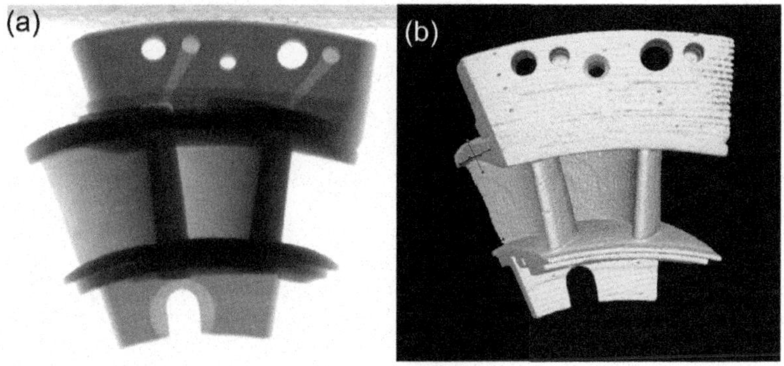

Fig. 1.3 The photograph of SiC/SiC twin guide vanes observed under XCT **a** Front view; and **b** Back view

In China, the National Key Lab of Thermostructure Composite Materials in Northwestern Polytechnical University designed and tested the SiC/SiC high-pressure turbine twin guide vanes (Liu et al. 2021), as shown in Fig. 1.3. Cyclic thermal shock tests were conducted at target temperatures of 1400, 1450, and 1480 °C in gas environment to analyze the damage mechanisms and failure modes. When the temperature increased from 1400 to 1480 °C, the spalling of the SiC coating occurred at the guide vane, and delamination occurred at the area near the trailing edge of the guide vane. Under the observation of X-ray computed tomography (XCT), no obvious damage inside of the guide vane was observed.

1.4.2 CMC Rotation Components

NASA Marshall Space Flight Center is committed to the development of turbopump C/SiC blisk for rocket engine, as shown in Fig. 1.4. The diameter of the C/SiC blisk was 19 cm and the thickness was about 2 cm. The test speed of the blisk approached $n = 25{,}000$ r/min at the high temperature of $T = 1093$ °C. During the test, the blades of the C/SiC blisk were cracked, but the blisk could still continue to work.

The Japan Aerospace Exploration Agency (JAXA) has developed a three-dimensional braided C/C turbine blisk with inner and outer diameters of 180 and 200 mm, respectively, and a rotational speed of 6,000 r/min (tip speed of 600 m/s) at room temperature. Rolls Royce proposes a method for the preparation of a split CMC turbine disk, which includes individual turbine blades, a turbine disk outer ring and a turbine disk body, and the turbine blades are fixed on the outer edge of the turbine disk outer ring. The General Electric (GE) Company for the first time introduced the SiC/SiC rotating parts into the harshest region of the aeroengine. The SiC/SiC low-pressure turbine blade was demonstrated on the F-414 turbofan engine for $N = 500$ cycles, as shown in Fig. 1.5. GE proposed a SiC/SiC turbine blisk design method with separate preparation and integrated assembly, where the center disk body is

Fig. 1.4 C/SiC turbine blisk developed by NASA Marshall Space Center

Fig. 1.5 The SiC/SiC turbine blade developed by GE company

made of metal and the SiC/SiC turbine blades are used in the high-temperature part of the outer edge.

In China, the National Key Lab of Thermostructure Composite Materials in Northwestern Polytechnical University designed and tested the SiC/SiC turbine blisk (Liu et al. 2023a), as shown in Fig. 1.6. The Spider Web Structure (SWS) SiC preform was used as the reinforcement in turbine blisk, and the BN interphase and SiC matrix were deposited on the surface of the SWS SiC fiber preform using the CVI method. The environmental barrier coatings (EBCs) were prepared on the surface of SiC/SiC turbine blisk using the atmospheric plasma spraying (APS) method. The NDE method was used to characterize the internal defects in SiC/SiC turbine blisk. The overspeed rotation test and engine test were conducted to check the turbine blisk. During the over rotation test, when the rotating speed reached $n = 104,166$ RPM, the blade in the turbine blisk broke; and when the rotating speed reached $n = 108,072$ RPM, the disk in the turbine blisk broke. During the engine bench test, low cyclic fatigue of $N = 994$ cyclic with maximum speed $n_{\max} = 60,000$ RPM and $N = 100$ cyclic with maximum speed $n_{\max} = 70,000$ RPM were completed. On January 1, 2022, the SiC/SiC turbine blisk successfully completed the first flight test in Zhuzhou,

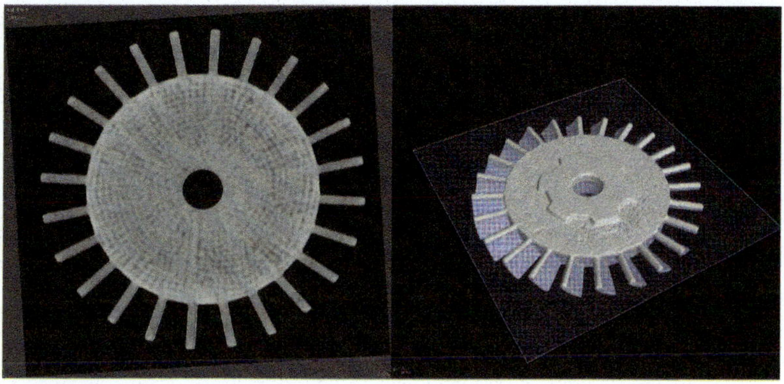

Fig. 1.6 CT scan of SiC/SiC turbine blisk

China, which is also the first flight test of the domestic SiC/SiC rotor assembly platform and verified the feasibility of SiC/SiC turbine rotor application in gas turbine engines.

1.5 Summary and Conclusions

In this chapter, the mechanical performance of CMCs, especially 2D SiC/SiC composite are given at first. The fabrication methods of CVI, PIP and MI were introduced, and the advantages and drawbacks for each fabrication method were analyzed. The working environments of the hot-section components of the aero-engine are introduced and the requirements for the CMCs applied on the hot-section components in aeroengines are given. The application cases for the stator and rotation hot-section components of aeroengines developed by different companies and institutes are introduced.

References

Bhatia T, Jarmon D, Shi J et al (2010) CMC combustor liner demonstration in a small helicopter engine. In: ASME Turbo Expo 2010: power for land, sea, and air. American Society of Mechanical Engineers Digital Collection, Glasgow, pp 509–513

Guo X, Wu J, Li J, Zeng Y, Huang X, Li L (2021) Damage monitoring of 2D SiC/SiC composites under monotonic and cyclic loading/unloading using acoustic emission and natural frequency. Ceram-Silik 65(2):125–131

Li L (2021) High temperature mechanical behavior of ceramic-matrix composites. Wiley-VCH, Weinheim, Germany

Li L (2022) High-temperature mechanical hysteresis behavior of ceramic-matrix composites. CRC Press, Taylor & Francis Group, London, UK

Li L (2023) Ceramic matrix composites: lifetime and strength prediction under static and stochastic loading. Elsevier, Oxford, UK
Li L (2024) Micromechanics of ceramic-matrix composites at elevated temperatures. Springer Nature, Singapore
Liu X, Guo X, Xu Y, Li LB, Zhu W et al (2021) Cyclic thermal shock damage behavior in CVI SiC/SiC high-pressure turbine twin guide vanes. Materials 14:6104
Liu X, Xu Y, Li J et al (2023a) Design, fabrication and testing of ceramic-matrix composite turbine blisk. Acta Materiae Compositae Sinica 40(3):1696–1706
Liu X, Zhao W, Guo H, Su Y, Li L, Liu C (2023b) XCT damage evaluation and analysis of SiC/SiC turbine guide vanes after thermal shocks. Int J Appl Ceram Technol 20:2919–2929

Chapter 2
High-Temperature Mechanical Properties of Ceramic-Matrix Composites

Abstract In this chapter, the high-temperature mechanical properties of chemical vapor infiltration (CVI) processed SiC/SiC composites with different fiber preforms, i.e., three-dimensional four-directional (3D4D) braided, two-dimensional plain-woven (2DPW) and stain woven (2DSW) lamination, are investigated at elevated temperature of 1400 °C. The high-temperature mechanical properties of SiC/SiC composites, e.g., tensile strength, compressional strength, interlaminar shear strength, in-plane shear strength, flexural strength, cyclic fatigue lifetime, and creep life are obtained from experimental data. The high-temperature mechanical properties can support the engineers to better design CMCs hot-section components.

Keywords SiC/SiC · Tensile · Compression, interlaminar shear · In-plane shear · Flexural · Fatigue · Creep · Mechanical properties

2.1 Introduction

Ceramic-matrix composites (CMCs) exhibit complex temperature-dependent thermomechanical properties, damage behavior and failure mechanisms (Li 2021, 2024, 2023, 2022). Guo and Kagawa (2002) investigated the tensile fracture behavior and tensile mechanical properties of polymer infiltration pyrolysis (PIP) processed two-dimensional plain-woven fabric SiC/SiC composite between 298 and 1400 K. The tensile strength increased slightly with increase in test temperature up to 1200 K; however, a large decrease in the strength was observed at 1400 K. The temperature depended tensile strength and fracture behavior of SiC/SiC composite was attributed to change of the in situ constituent properties with temperature. Li (2017, 2019, 2020a) analyzed the temperature-dependent matrix cracking and evolution in CMCs, and developed micromechanical approaches to predict the first matrix cracking stress,

X. Liu and L. Li, *Design, Fabrication and Testing of Aeroengine Ceramic-Matrix Composite Components*, Advanced Ceramics and Composites 7,
https://doi.org/10.1007/978-981-97-6109-8_2

and proportional limit stress in CMCs at elevated temperatures. Li (2020b) investigated the temperature and time-dependent tensile behavior of CMCs considering interface and fiber oxidation, and developed a micromechanical approach to predict the tensile stress–strain curves of SiC/SiC composites for different damage stages at elevated temperatures. Under cyclic fatigue loading, the temperature affects the fatigue damage mechanisms of CMCs, e.g., matrix cracking evolution (Li 2013), interface shear stress degradation (Li 2014), and fibers failure (Li 2016). Dalmaz et al. (1998) investigated the mechanical behavior and damage progression of the 2.5D woven C/SiC composite submitted to cyclic fatigue loading at elevated temperature. It was found that the impact of cyclic fatigue is related to shear strength at the fiber/matrix interface. Li (2015a, b) developed fatigue hysteresis dissipated energy-based features for damage monitoring and life prediction of fiber-reinforced CMCs submitted to cyclic fatigue loading. Damage mechanisms, including matrix cracking, interface debonding and sliding, and fiber damage and fracture, have been monitored. Under tensile creep loading, the strain rates of 2.5D woven C/SiC achieved in the steady state was between 2×10^{-9}/s (110 MPa, 1000 °C) and 2×10^{-8}/s (220 MPa 1400 °C) (Boitier et al. 1997). Five different types of cracks occur under creep loading, including transverse cracks in the transverse yarns, transverse cracks in the longitudinal yarns, longitudinal cracks in the transverse yarns, longitudinal cracks in the longitudinal yarns, and interplay cracks (Boitier et al. 1998). Li et al. (2022) developed a micromechanical approach to predict the damage evolution in CMCs subjected to creep-fatigue loading at elevated temperatures. Synergistic effects of creep fatigue peak stress level and damage in the matrix and the interface on time-dependent damage evolution in C/SiC composites were analyzed.

In this chapter, the high-temperature mechanical properties of chemical vapor infiltration (CVI) processed SiC/SiC composites with different fiber preforms, i.e., three-dimensional four-directional (3D4D) braided, two-dimensional plain-woven (2DPW) and stain woven (2DSW) lamination, are investigated at elevated temperature of 1400 °C. The composites' tensile strength, compressional strength, interlaminar shear strength, in-plane shear strength, flexural strength, cyclic fatigue lifetime, and creep life of SiC/SiC composites are experimentally obtained. The experimental mechanical properties can support the engineers to better design CMCs hot-section components.

2.2 High-Temperature Tensile Mechanical Properties

Tensile tests of chemical vapor infiltration (CVI) processed three-dimensional four-directional (3D4D) braided, two-dimensional plain woven (2DPW) and satin woven (2DSW) laminated CVI SiC/SiC composites were conducted at a high temperature of 1400 °C, and their ultimate tensile strength (UTS), 0.05% proportional limit stress (PLS), fracture strain (FS), and tensile stress–strain curves were all obtained. The

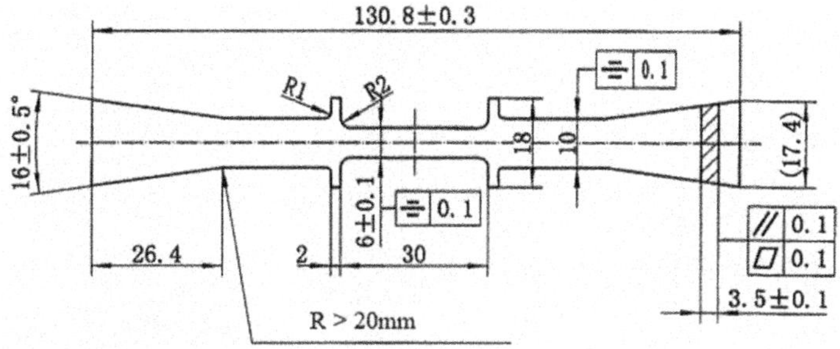

Fig. 2.1 Shape and geometry parameters of high-temperature tensile SiC/SiC specimen (unit: mm)

shape and geometry of high-temperature tensile SiC/SiC specimen are shown in Fig. 2.1. The tensile experiments were conducted on INSTRON 8801 with the ultra-high temperature heating system, as shown in Fig. 2.2. The test environment is vacuum, i.e., 10^{-3} Pa. Use S-type thermocouple to test the temperature, and the distance between the thermocouple temperature measuring end and the specimen is about 1–3 mm. The heating rate of the specimen is 20 °C/min, and hold for 10 min after reaching the test temperature. Under tensile loading, the loading rate was 0.5 mm/min. Figure 2.3 shows the typical tensile stress–strain curves of 3D4D, 2DPW and 2DSW SiC/SiC composites at 1400 °C. The tensile mechanical properties of 3D4D, 2DPW and 2DSW SiC/SiC composites are listed in Table 2.1. The tensile strength and the fracture strain of 2DPW SiC/SiC composite are the highest with the average value of $\sigma_{\text{uts}} = 285$ MPa; and the proportional limit stress of 2DSW SiC/SiC composite is the highest with the average value of $\sigma_{\text{pls}} = 175$ MPa; and the tensile strength and the elastic modulus of 3D4D SiC/SiC composite is the lowest. By observing the macroscopic fracture surfaces of three types of SiC/SiC composites under SEM, it can be found that the fracture surfaces of 2DPW and 2DSW SiC/SiC composites are flat, with obvious transverse cracks, and the 2DSW SiC/SiC exhibits very short fibers pulling out, and the fracture positions are all located in the measuring gauge range; and the fracture surfaces of the 3D4D SiC/SiC composites exhibit obvious pulling out of the fiber bundles, and the angle between the fracture direction and loading direction is approximately 45°, which depends on the fiber preforms.

2.3 High-Temperature Compression Mechanical Properties

The compression test of chemical vapor infiltration (CVI) processed 2DPW SiC/SiC composite at 1400 °C were conducted under the displacement control with the loading rate of 0.5 mm/min. The shape and geometry parameters of the high-temperature compression SiC/SiC specimens are shown in Fig. 2.4. Three specimens are conducted for the high-temperature compression tests. The typical compression

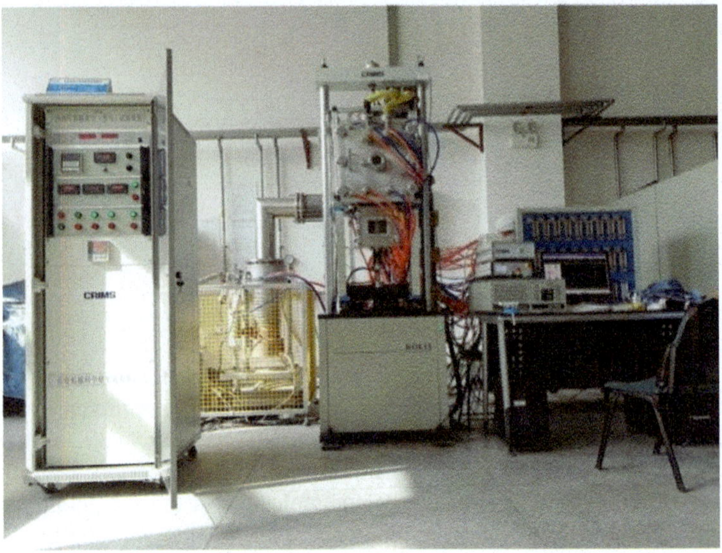

Fig. 2.2 High temperature tensile experiment system

load–displacement curve is shown in Fig. 2.5. The compressive strength of SiC/SiC composites is $\sigma_{cs} = 263.7 \pm 34.5$ MPa with a large dispersion of compressive strength values. The high-temperature compression macroscopic fracture morphology is shown in Fig. 2.6. It can be seen that the fracture mode is a typical pure compression damage mode, and the cross-section is about 30° angle with the loading direction; the fracture is relatively flat, and some of the warp fibers bundles are broken due to the longitudinal instability in the compression experimental process, and some of the fibers are split, and there also exists matrix cracking on the fracture surface.

2.4 High-Temperature Interlaminar Shear Mechanical Properties

The interlaminar shear test of chemical vapor infiltration (CVI) processed 2DPW SiC/SiC composite at 1400 °C were conducted under the displacement control with the loading rate of 0.5 mm/min. The shape and geometry parameters of the high-temperature interlaminar SiC/SiC specimens are shown in Fig. 2.7. The interlaminar shear strength σ_{ILSS} is,

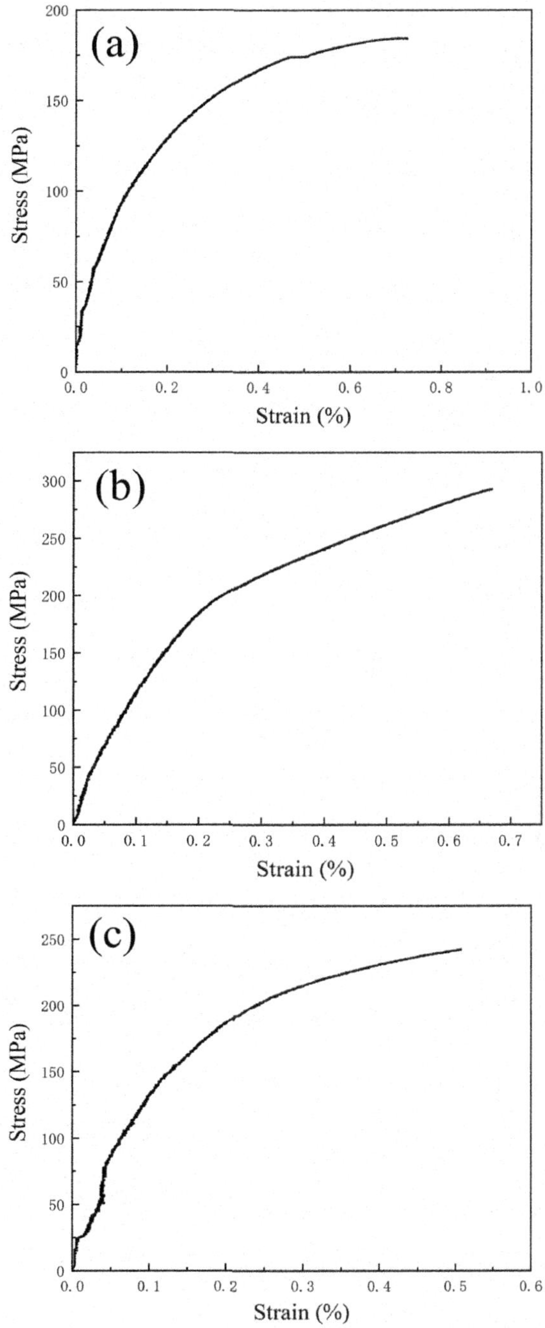

Fig. 2.3 Typical tensile stress–strain curves of **a** 3D4D, **b** 2DPW and **c** 2DSW SiC/SiC composites at 1400 °C

Table 2.1 Tensile mechanical properties of SiC/SiC composite at 1400 °C

Sample number	Modulus (GPa)	Proportional limit stress (MPa)	Ultimate tensile strength (MPa)	Fracture strain (%)
3D4D-01	130	90	185	0.72
3D4D-02	230	108	210	0.69
3D4D-03	164	–	112	–
Average value	180	99	198	0.71
Standard deviation	41.57	–	41.57	–
2DPW-01	130	150	301	1.26
2DPW-02	150	145	293	0.67
2DPW-03	–	–	260	0.54
Average value	140	147.5	285	0.82
Standard deviation	–	–	17.74	0.31
2DSW-01	200	190	242	0.51
2DSW-02	–	–	253	0.45
2DSW-03	220	160	248	0.59
Average value	210	175	248	0.52
Standard deviation	–	–	4.50	0.06

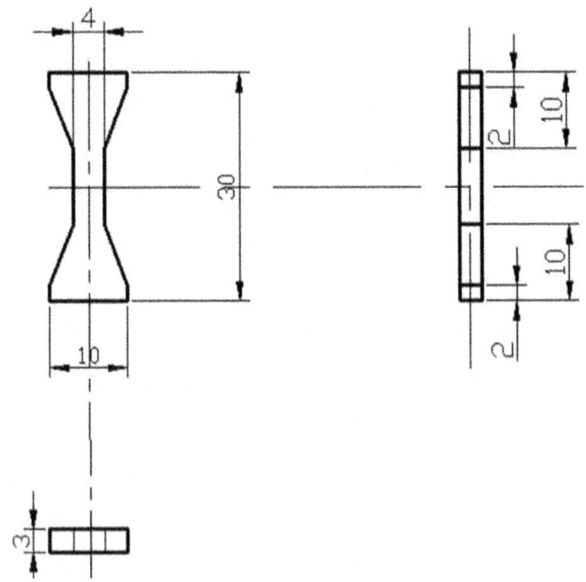

Fig. 2.4 Shape and geometry parameters of high-temperature compression 2DPW SiC/SiC specimen (unit: mm)

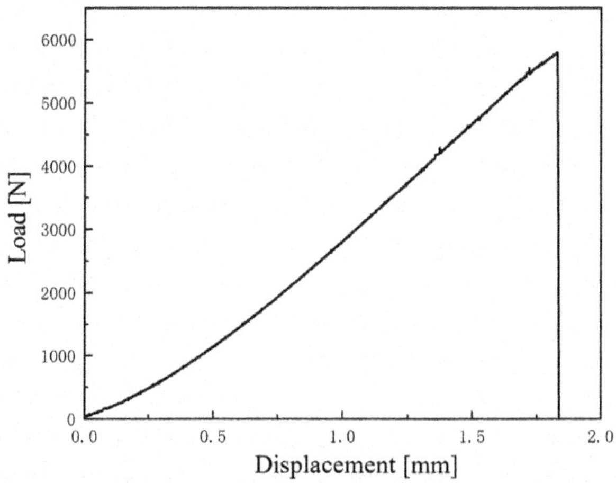

Fig. 2.5 Compression load–displacement curves of 2DPW SiC/SiC composite at 1400 °C

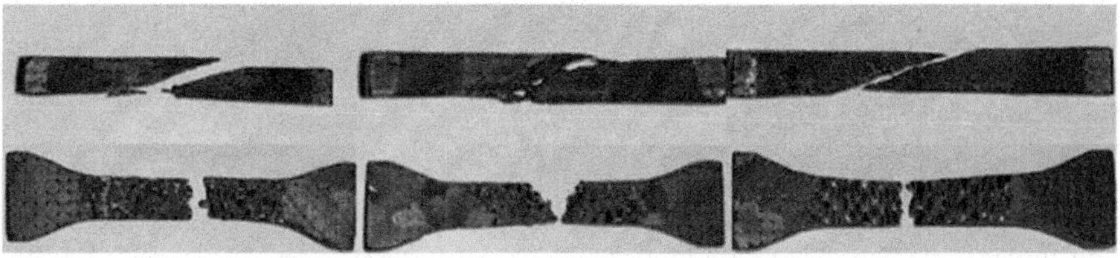

Fig. 2.6 High-temperature compression macroscopic fracture morphology of 2DPW SiC/SiC specimens

$$\sigma_{\mathrm{ILSS}} = \frac{F_{\mathrm{m}}}{bl} \tag{2.1}$$

where F_m is the peak load (unit: N), b is the specimen width (unit: mm), and l is the distance between notches (unit: mm).

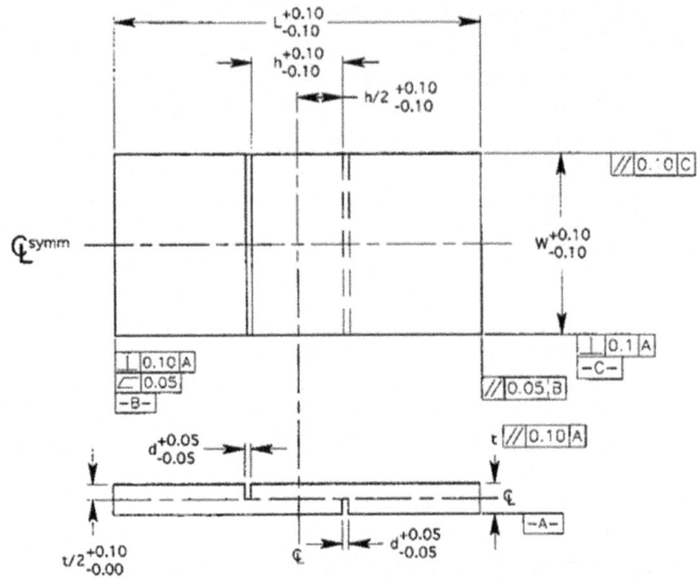

Fig. 2.7 Shape and geometry parameters of the high-temperature interlaminar shear 2DPW SiC/SiC specimens (unit: mm)

Four specimens are conducted for the high-temperature interlaminar shear tests. The typical interlaminar shear load–displacement curve is shown in Fig. 2.8. The interlaminar shear strength of 2DPW SiC/SiC composite is $\sigma_{\mathrm{ILSS}} = 29.9 \pm 4.9$ MPa. Figure 2.9 shows the high-temperature interlaminar shear macroscopic fracture morphology. Obvious fiber distribution can be observed at the shear fracture surface of the interlaminar shear specimen, and the fracture mode is a standard interlaminar shear fracture with transverse fiber shear fracture and flat fracture.

2.5 High-Temperature In-Plane Shear Mechanical Properties

The in-plane shear test of chemical vapor infiltration (CVI) processed 2DPW SiC/SiC composite at 1400 °C are conducted under the displacement control with the loading rate of 0.5 mm/min. The schematic of the high-temperature in-plane shear SiC/SiC specimen is shown in Fig. 2.10. The in-plane shear strength σ_{IPS} is,

$$\sigma_{\mathrm{IPS}} = \frac{F_{\mathrm{m}}}{lh} \tag{2.2}$$

where F_{m} is the peak load (unit: N), l is the distance between notches (unit: mm), and h is the specimen width (unit: mm).

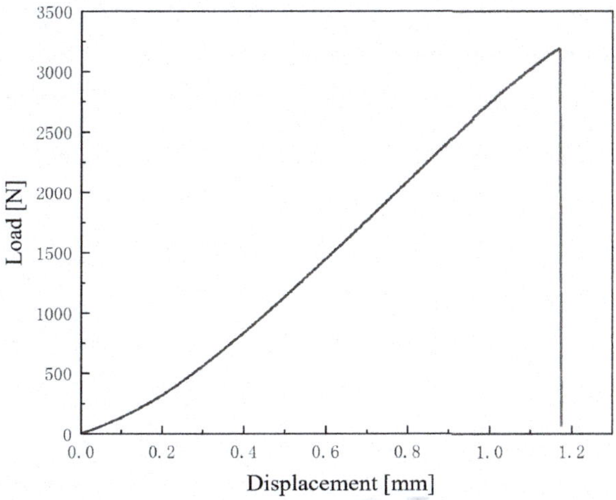

Fig. 2.8 Typical interlaminar shear load–displacement curves of 2DPW SiC/SiC composite at 1400 °C

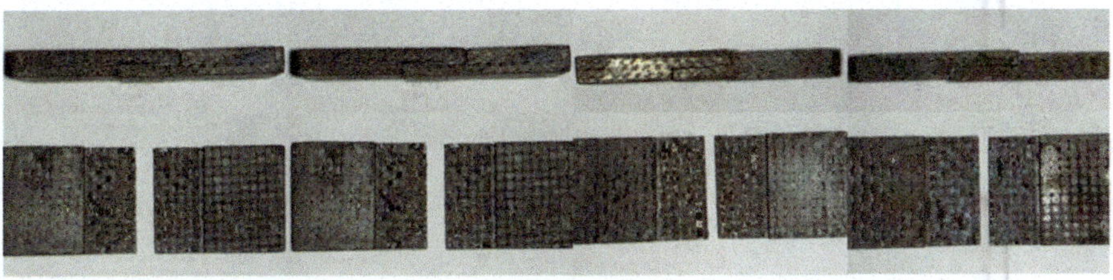

Fig. 2.9 High-temperature interlaminar shear macroscopic fracture morphology of 2DPW SiC/SiC specimens

Fig. 2.10 Schematic of the high-temperature in-plane shear 2DPW SiC/SiC specimens

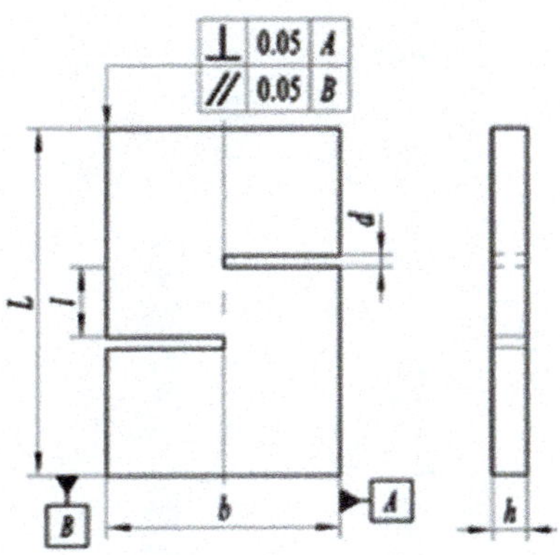

Five specimens are conducted for the in-plane shear tests at 1400 °C. Typical in-plane shear load–displacement curve is shown in Fig. 2.11. The in-plane shear strength of PW SiC/SiC composite is $\sigma_{IPS} = 207.4 \pm 31.9$ MPa. Figure 2.12 shows the high-temperature in-plane shear macroscopic fracture morphology. The fracture surface is smooth and flat without fiber pulling out, and there exists pores between inter-fibers.

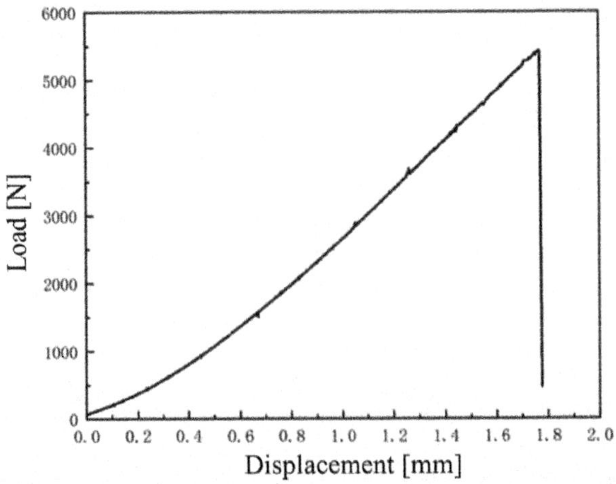

Fig. 2.11 Typical in-plane shear load–displacement curve of 2DPW SiC/SiC composite at 1400 °C

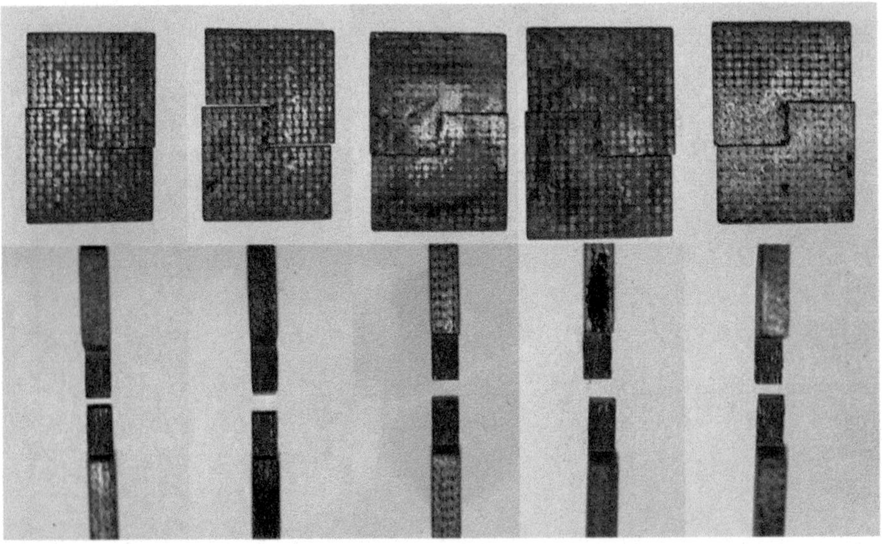

Fig. 2.12 High-temperature in-plane shear macroscopic fracture morphology of 2DPW SiC/SiC specimens

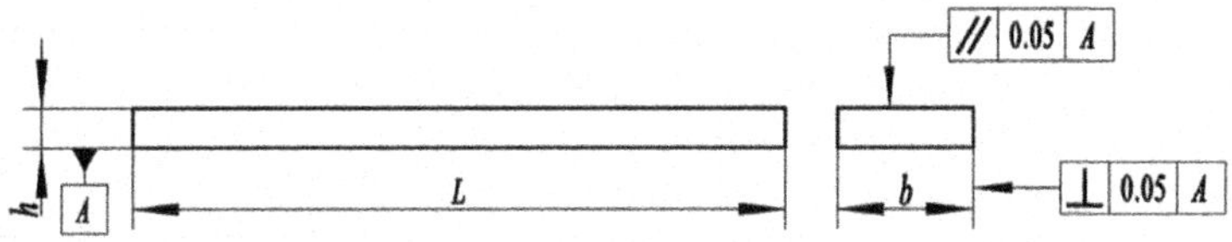

Fig. 2.13 Schematic of the high-temperature flexural SiC/SiC specimens

2.6 High-Temperature Flexural Mechanical Properties

The flexural tests of chemical vapor infiltration (CVI) processed 2DPW, 2DSW and 3D4D SiC/SiC composites at 1400 °C are conducted under the displacement control with the loading rate of 1.0 mm/min. The schematic of the high-temperature flexural SiC/SiC specimens is shown in Fig. 2.13. The flexural strength σ_{FS} is,

$$\sigma_{\mathrm{FS}} = \frac{3F_{\mathrm{m}}L_{\mathrm{u}}}{2bh^2} \tag{2.3}$$

where F_{m} is the peak load, L_{u} is the span length, b is the specimen width, and h is the height of the specimen.

Figure 2.14 shows the typical flexural tests of 2DPW, 2DSW, and 3D4D SiC/SiC composites at 1400 °C. The flexural strength of 2DPW, 2DSW and 3D4D SiC/SiC composites are $\sigma_{\mathrm{FS}} = 327.06 \pm 28.64$, 417.65 ± 16.23, and 419.83 ± 27.02 MPa, respectively. Table 2.2 lists the flexural strength of 2DPW, 2DSW, and 3D4D SiC/SiC composites at 1400 °C. Figure 2.15 shows the high-temperature flexural macroscopic fracture morphology. The 2DSW SiC/SiC composite has less deformation, the 3D4D SiC/SiC composite has the least deformation, while the 2DPW SiC/SiC composite has the greatest deformation and the greatest degree of fracture splitting.

2.7 High-Temperature Fatigue Mechanical Properties

The fatigue tests of chemical vapor infiltration (CVI) processed 2DPW, 2DSW and 3D4D SiC/SiC composites at 1400 °C are conducted under the load control with the stress ratio of 0.1 and loading frequency of 20 Hz. The fatigue limit is defined as 10^6 cycles. The schematic of the high-temperature fatigue SiC/SiC specimens is shown in Fig. 2.16. The high-temperature fatigue experimental results of 2DPW, 2DSW and 3D4D SiC/SiC composite are listed in Table 2.3. For 2DPW SiC/SiC composites, the fatigue peak stresses are $\sigma_{\mathrm{max}} = 70$, 140, 160 and 180 MPa with the corresponding cycles of 10^6, 10^6, 10^6 and 53,651, respectively; for 2DSW SiC/SiC

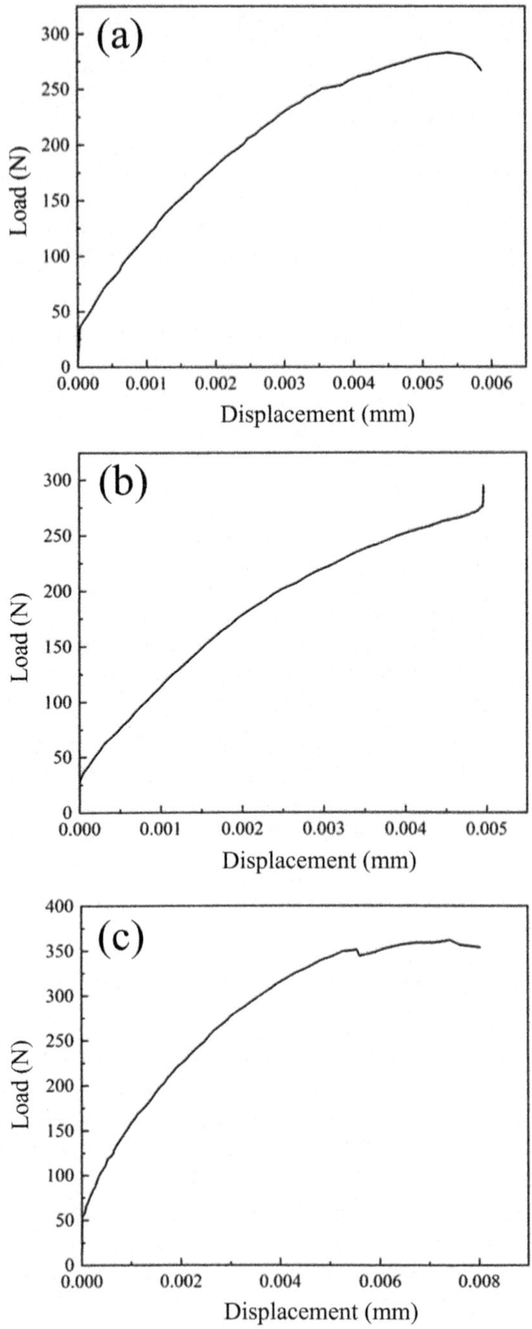

Fig. 2.14 Flexural load–displacement curves of **a** 2DPW, **b** 2DSW, and **c** 3D4D SiC/SiC composites at 1400 °C

Table 2.2 Flexural strength of SiC/SiC composites at 1400 °C

Sample number	b (mm)	h (mm)	Flexural strength (MPa)
2DPW-01	3.60	5.26	295.35
2DPW-02	3.74	5.20	300.47
2DPW-03	3.74	5.20	341.62
2DPW-04	3.64	5.20	324.24
2DPW-05	3.56	5.20	373.63
Average value	–	–	327.06
Standard deviation	–	–	28.64
2DSW-01	3.40	5.20	441.09
2DSW-02	3.64	5.20	401.56
2DSW-03	3.44	5.20	397.30
2DSW-04	3.54	5.20	426.37
2DSW-05	3.44	5.20	421.93
Average value	–	–	417.65
Standard deviation	–	–	16.23
3D4D-01	3.90	5.50	389.47
3D4D-02	3.84	5.20	449.64
3D4D-03	3.84	5.20	443.68
3D4D-04	3.84	5.30	396.56
Average value	–	–	419.83
Standard deviation	–	–	27.02

composite, the fatigue peak stresses are $\sigma_{max} = 100$, 160, 170 and 200 MPa with the corresponding cycles of 10^6, 15,694, 41,763 and 8, respectively; and for 3D4D SiC/SiC composite, the fatigue peak stresses are $\sigma_{max} = 70$, 120, 140 and 160 MPa with the corresponding cycles of 10^6, 10^6, 10^6 and 74,361.

Figure 2.17 shows the macro fracture morphology of SiC/SiC composites after fatigue fracture. It can be seen that the fracture morphology of the 2DPW and 2DSW SiC/SiC composite is relatively smooth, accompanied by a certain number of fibers pulling out; while the fracture of the 3D4D SiC/SiC composite has obvious fiber bundles pulling out, the fracture morphology is more rough compared to the 2DPW and 2DSW SiC/SiC composite.

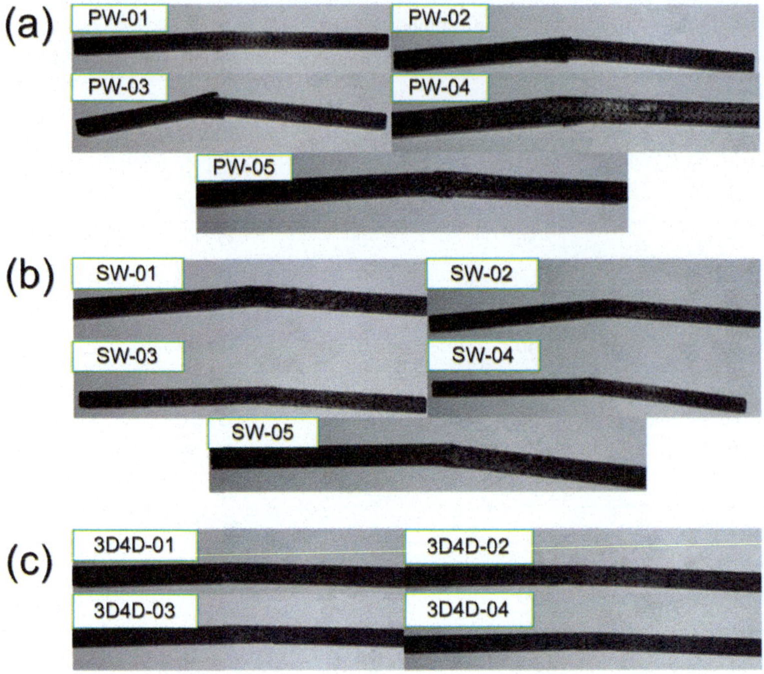

Fig. 2.15 High-temperature in-plane shear macroscopic fracture morphology of **a** 2DPW, **b** 2DSW, and **c** 3D4D SiC/SiC specimens

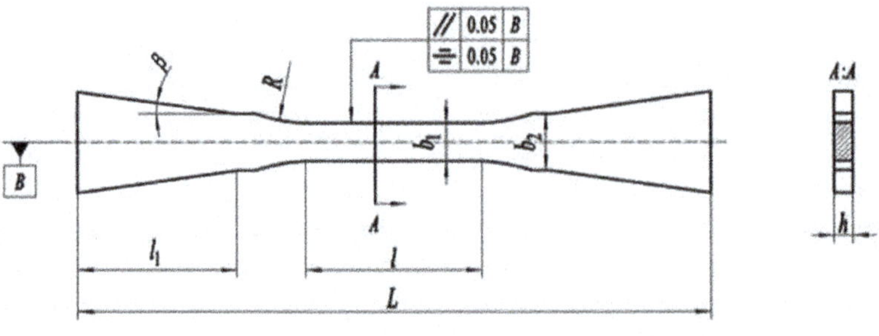

Fig. 2.16 Schematic of the high-temperature fatigue SiC/SiC specimens

2.8 High-Temperature Creep Mechanical Properties

The creep tests of chemical vapor infiltration (CVI) processed 2DPW SiC/SiC composite at T = 1400, 1410 and 1430 °C are conducted under the stress of σ = 80 MPa. The schematic of creep specimen is shown in Fig. 2.18. The creep strain curves of 2DPW SiC/SiC composite at T = 1400, 1410 and 1430 °C under σ = 80 MPa are shown in Fig. 2.19. The creep experimental results are listed in Table 2.4.

Table 2.3 High temperature fatigue experiment results of 2DPW, 2DSW and 3D4D SiC/SiC composites

Sample number	Width (mm)	Thickness (mm)	σ_{max} (MPa)	N
2DPW-01	5.60	3.64	70	10^6
2DPW-02	5.72	3.68	140	10^6
2DPW-03	6.06	3.62	160	10^6
2DPW-04	5.78	3.58	180	53,651
2DSW-01	6.13	3.52	100	10^6
2DSW-02	6.04	3.52	160	15,694
2DSW-03	6.02	3.50	170	41,763
2DSW-04	6.06	3.52	200	8
3D4D-01	6.14	3.65	70	10^6
3D4D-02	6.24	3.68	120	10^6
3D4D-03	6.26	3.74	140	10^6
3D4D-04	6.26	3.64	160	74,361

- At $T = 1400$ °C under $\sigma = 80$ MPa, the creep duration is 16.1 h with the minimum creep rate of 4.49×10^{-8}/s, and creep fracture strain of 0.647%.
- At $T = 1410$ °C under $\sigma = 80$ MPa, the creep duration is 13.7 h with the minimum creep rate of 6.10×10^{-7}/s, and creep fracture strain of 4.596%.
- At $T = 1430$ °C under $\sigma = 80$ MPa, the creep duration is 10.1 h with the minimum creep rate of 9.52×10^{-7}/s, and creep fracture strain of 4.907%.

Figure 2.20 shows the macro fracture morphology of SiC/SiC composites after creep fracture. There exists obvious matrix cracks and pullout fibers at the fracture surface.

2.9 Summary and Conclusions

In this chapter, the high-temperature mechanical properties of chemical vapor infiltration (CVI) processed SiC/SiC composites with different fiber preforms, i.e., three-dimensional four-directional (3D4D) braided, two-dimensional plain-woven (2DPW) and stain woven (2DSW) lamination, are investigated at elevated temperature of 1400 °C. The composites' tensile strength, compressional strength, interlaminar shear strength, in-plane shear strength, flexural strength, cyclic fatigue lifetime and creep life of SiC/SiC composites are experimentally obtained.

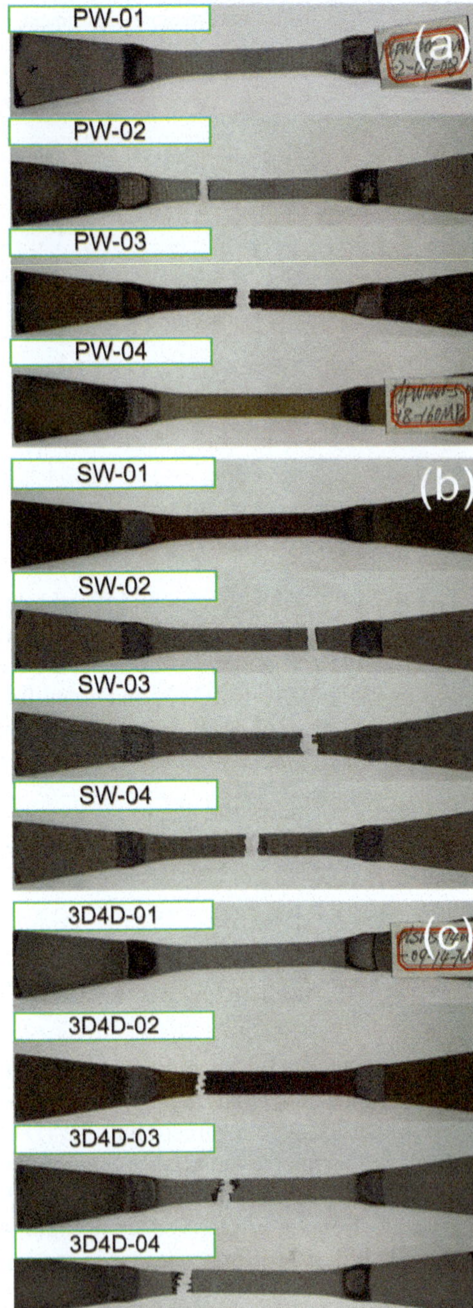

Fig. 2.17 High-temperature fatigue macroscopic fracture morphology of **a** 2DPW, **b** 2DSW, and **c** 3D4D SiC/SiC specimens

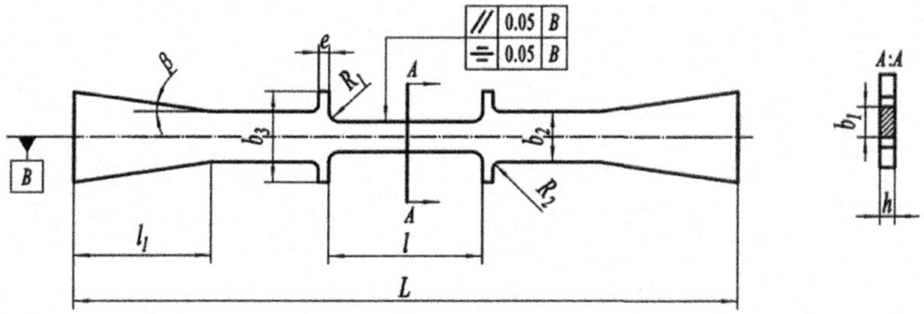

Fig. 2.18 Schematic of the high-temperature creep SiC/SiC specimens

- For the high-temperature tensile properties, the tensile strength and the fracture strain of 2DPW SiC/SiC composite are the highest with the average value of $\sigma_{\text{uts}} = 285$ MPa; and the proportional limit stress of 2DSW SiC/SiC composite is the highest with the average value of $\sigma_{\text{pls}} = 175$ MPa; and the tensile strength and the elastic modulus of 3D4D SiC/SiC composite are the lowest.
- For the high-temperature compression properties, the compressive strength of 2DPW SiC/SiC composites is $\sigma_{\text{cs}} = 263.7 \pm 34.5$ MPa with a large dispersion of compressive strength values.
- For the high-temperature interlaminar shear properties, the interlaminar shear strength of 2DPW SiC/SiC composite is $\sigma_{\text{ILSS}} = 29.9 \pm 4.9$ MPa.
- For the high-temperature in-plane shear properties, the in-plane shear strength of PW SiC/SiC composite is $\sigma_{\text{IPS}} = 207.4 \pm 31.9$ MPa.
- For the high-temperature flexural properties, The flexural strength of 2DPW, 2DSW and 3D4D SiC/SiC composites are $\sigma_{\text{FS}} = 327.06 \pm 28.64$, 417.65 ± 16.23, and 419.83 ± 27.02 MPa, respectively.

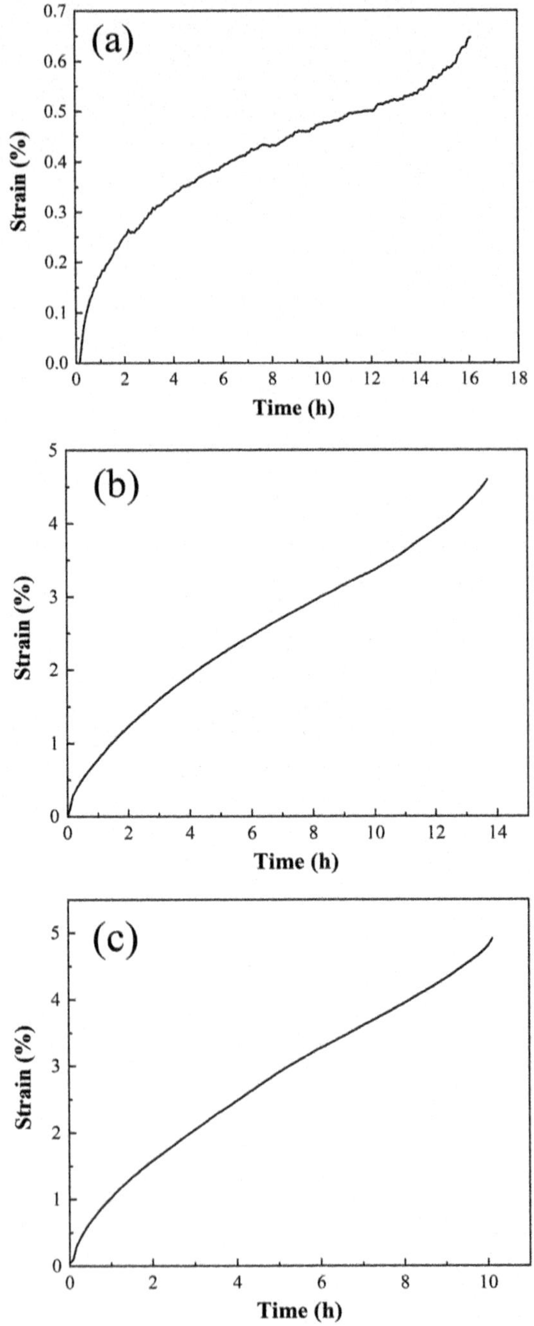

Fig. 2.19 Creep strain curves of 2DPW SiC/SiC composite under σ = 80 MPa at **a** 1400 °C; **b** 1410 °C; and **c** 1430 °C

Table 2.4 High temperature creep experiment results of 2DPW SiC/SiC composite

Temperature (°C)	Stress (MPa)	Creep life (h)	Minimum creep rate (s^{-1})	Creep strain (%)
1400	80	16.1	4.49×10^{-8}	0.647
1410		13.7	6.10×10^{-7}	4.596
1430		10.1	9.52×10^{-7}	4.907

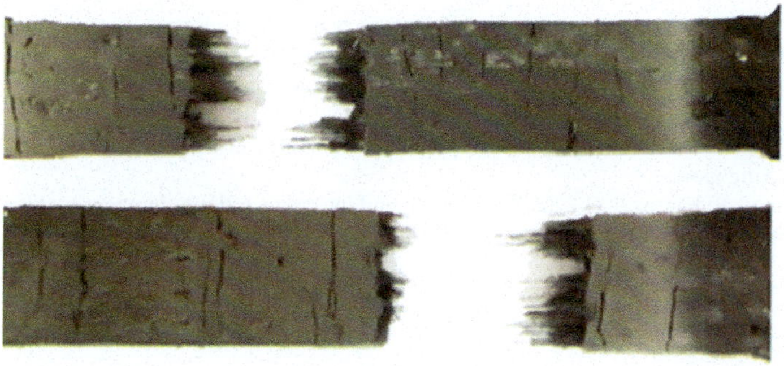

Fig. 2.20 High-temperature creep macroscopic fracture morphology of 2DPW SiC/SiC composite

References

Boitier G, Vicens J, Chermant JL (1997) Tensile creep results on a C_f-SiC composite. Scr Mater 37:1923–1997

Boitier G, Vicens J, Chermant JL (1998) Microstructure and creep of 2.5D C_f-SiC composites. J Eur Ceram Soc 18:1835–1843

Dalmaz A, Reynaud P, Rouby D, Fantozzi, Abbe F (1998) Mechanical behavior and damage development during cyclic fatigue at high-temperature of a 2.5D carbon/sic composite. Compos Sci Technol 58:693−699

Guo S, Kagawa Y (2002) Tensile fracture behavior of continuous SiC fiber-reinforced SiC matrix composites at elevated temperatures and correlation to in situ constituent properties. J Eur Ceram Soc 22:2349–2356

Li L (2013) Fatigue hysteresis behavior of cross-ply C/SiC ceramic matrix composites at room and elevated temperatures. Mater Sci Eng A 586:160–170

Li L (2014) Assessment of the interfacial properties from fatigue hysteresis loss energy in ceramic-matrix composites with different fiber preforms at room and elevated temperatures. Mater Sci Eng A 613:17–36

Li L (2015a) A hysteresis dissipated energy-based parameter for damage monitoring of carbon fiber-reinforced ceramic-matrix composites under fatigue loading. Mater Sci Eng A 634:188–201

Li L (2015b) A hysteresis dissipated energy-based damage parameter for life prediction of carbon fiber-reinforced ceramic-matrix composites under fatigue loading. Compos B 82:108–128

Li L (2016) Damage monitoring of unidirectional C/SiC ceramic-matrix composite under cyclic fatigue loading using a hysteresis loss energy-based damage parameter at room and elevated temperatures. Appl Compos Mater 23:357–374

Li L (2017) Modeling matrix cracking of fiber-reinforced ceramic-matrix composites under oxidation environment at elevated temperature. Theoret Appl Fract Mech 87:110–119

Li L (2019) Time-dependent damage and fracture of fiber-reinforced ceramic-matrix composites at elevated temperatures. Compos Interfaces 26(11):963–988

Li L (2020a) Synergistic effects of temperature and time on proportional limit stress of silicon carbide fiber-reinforced ceramic-matrix composites. Compos Interfaces 27(4):341–353

Li L (2020b) A time-dependent tensile constitutive model for long-fiber-reinforced unidirectional ceramic-matrix minicomposites considering interface and fiber oxidation. Int J Damage Mech 29(7):1138–1166

Li L (2021) High temperature mechanical behavior of ceramic-matrix composites. Wiley-VCH, Weinheim, Germany

Li L (2022) High-temperature mechanical hysteresis behavior of ceramic-matrix composites. CRC Press, Taylor & Francis Group, London, UK

Li L (2023) Ceramic matrix composites: lifetime and strength prediction under static and stochastic loading. Elsevier, Oxford, UK

Li L (2024) Micromechanics of ceramic-matrix composites at elevated temperatures. Springer Nature, Singapore

Li L, Reynaud P, Fantozzi G (2022) Time-dependent creep fatigue damage evolution in C/SiC composite: theory and analytical prediction. Ceram Int 48:20731–20742

Chapter 3
Design, Fabrication and Testing of CMC High-Pressure Turbine Guide Vane

Abstract In this chapter, the design, fabrication and testing of CMC high-pressure single/twin turbine guide vanes were investigated. The service performance analysis of the turbine guide vane was at first analyzed. The SiC/SiC high-pressure turbine single/twin guide vanes were fabricated using the chemical vapor infiltration (CVI) method. Thermal shock experiments of SiC/SiC high-pressure turbine guide vanes were carried out on the third-generation turbine service environment simulation device. For single turbine guide vane, the thermal shock tests between 350 and 1350 °C in gas environment were conducted; and for twin turbine guide vane, cyclic thermal shock tests at different target temperatures (i.e., 1400, 1450, and 1480 °C) in gas environment were conducted to investigate the damage mechanisms and failure modes. During thermal shock tests, the surface temperature field of the SiC/SiC guide vane was analyzed. The damage characteristics of the single/twin turbine guide vane after experiencing different thermal shocks were discussed.

Keywords SiC/SiC · Turbine guide vane · Thermal shock test · Temperature field

3.1 Introduction

Fiber reinforced ceramic-matrix composite (CMC) materials, such as SiC/SiC, have higher temperature capabilities than conventional metallic materials used in the first stage of the High Pressure Turbine (HPT) (Li 2024, 2023, 2022, 2021a; Liu et al. 2021). Using CMC vanes in the first stage of the HPT provides the greatest benefit in terms of reduced coolant and reduced NO_x and CO_2 emissions. In USA, under NASA's Enabling Propulsion Materials and Ultra Efficient Engine Technologies programs, the SiC/SiC turbine guide vane was developed (Verrilli et al. 2004). Three guide vanes with Haynes 188 superalloy vanes were tested using NASA's High Pressure Burner Rig (HPBR) for 50 h of steady state operation and 102 thermal cycles. The minimum gas temperature ranged from 900 to 1050 °C and the maximum temperature ranged from 940 to 1440 °C. After 50 h testing, about 26% of the total coating surface area had spalled. After thermal cycle tests, melting of the superalloy

X. Liu and L. Li, *Design, Fabrication and Testing of Aeroengine Ceramic-Matrix Composite Components*, Advanced Ceramics and Composites 7,
https://doi.org/10.1007/978-981-97-6109-8_3

vanes occurred, however, the SiC/SiC vanes were intact with deposites on the EBC surface. In Japan, the IHI Corporation developed the SiC/SiC turbine guide vane and tested in the IHI IM270 industrial gas turbine engine (Fumiaki et al. 2017). Cyclic thermal shock test for SiC/SiC vane was conducted by burner rig. The maximum temperature was set to 1200 °C. 1000 thermal cycles were conducted with 6 min for each thermal cycle. After thermal cyclic tests, the cracks were found at leading edge due to the machining. Engine test was conducted on the IHI IM270 with four SiC/SiC vanes installed into the 1st stage turbine nozzle assembly. The engine test conducted for 413 h with 48 starts and normal stops. The inlet gas temperature was about 1050 °C at steady state.

In this chapter, the design, fabrication and testing of CMC high-pressure single/twin turbine guide vanes were investigated. The service performance analysis of the single/twin turbine guide vane was at first analyzed. The SiC/SiC high-pressure turbine single/twin guide vanes were fabricated, and the thermal shock experiments were conducted to investigate the damage mechanisms and failure modes. For single turbine guide vane, the thermal shock tests between 350 and 1350 °C in gas environment were conducted; and for twin turbine guide vane, cyclic thermal shock tests at different target temperatures (i.e., 1400, 1450, and 1480 °C) in gas environment were conducted. During thermal shock tests, the surface temperature field of the SiC/SiC guide vane was analyzed. After thermal shock tests, the damage characteristics of the single/twin turbine guide vane were discussed.

3.2 Service Performance Analysis of Turbine Guide Vane

The turbine guide vanes are located in the rear section of the combustion chamber, mainly converting the thermal energy of the high temperature gas into kinetic energy, so that the direction of the gas flow meets the working requirements. Due to the direct contact with gas, the service environment of the guide vane is extremely harsh, which puts forward high requirements on the temperature resistance of the material. With the continuous increase of turbine inlet temperature, the service environment of the guide vane further deteriorates.

(1) High temperatures: In steady state operation, the turbine guide vanes will be directly heated to extremely high temperatures by the high temperature gas. For example, the turbine inlet temperature of the F119 engine has already reached 1800 K, and the average temperature of the turbine inlet of the future engine will exceed 2000 K.
(2) Thermal stress load: under the action of high temperature gas and cooling air, different parts of the guide vane are not uniformly heated, resulting in the existence of a large thermal stress; in addition, when the engine is in different working conditions, e.g., starting and stopping, acceleration and deceleration, cruising, etc., the rapid heating, cooling or high and low temperature cycling of the gas is very easy to cause large amplitude of temperature change of the guide

vane, resulting in the existence of a large temperature gradient and thermal stress load inside the guide vane.

(3) Pneumatic-vibration load: the guiding vane needs to withstand the aerodynamic force of high-temperature and high-pressure gas in the working process. Due to the fluctuation of gas flow, changes in the working condition, etc., it is easy to cause additional vibration load on the guide vanes.
(4) Particle erosion: airborne pollutants, gravel and other hard particles or engine flaking debris, will be existed in the gas flow, and cause high-speed impact to the surface of the guide vane, leading to the damages on the surface of the environmental barrier coating or the guide vanes.
(5) High-temperature corrosion: Some sand or volcanic ash in the air will easily be deposited on the surface of the guide vane after melting in the high-temperature gas heating, which will increase the roughness of the surface of the guide vane, and thus reduce its aerodynamic performance; the molten deposits will gradually penetrate into the coating or the guide vane, and may react with it chemically and lead to the failure of the coating or the SiC matrix. In addition, surface deposits tend to block the air film holes, leading to insufficient cooling and shortening the guide vane life.

3.3 Design and Fabrication of CMC Turbine Guide Vanes

The SiC/SiC composite high-pressure turbine single/twin guide vanes were fabricated using the chemical vapor infiltration (CVI) method. The guide vanes were formed by 2D lamination. The process of stacking and layering plain-woven cloth and Z-direction puncture and suture was adopted. The fiber's volume fraction is 40–45%. The SiC fiber's diameter was 14–16 μm and the density is approximately 2.75 g/cm^3. To protect the SiC fiber and adjust the thermal misfit between the fiber and the matrix, the PyC interphase was deposited on the surface of the fiber using the chemical vapor deposition (CVD) method. The CVD temperature was approximately 900–1100 °C using the Ar as the protective gas. The deposition time is approximately 20–50 h and the thickness of the interphase is approximately 300–500 μm. After deposition of the interface, the SiC matrix was deposited on the fiber's preform using the CVI method at approximately 1100–1400 °C. The deposition furnace was vacuumized to 20–50 kPa, 60–100 L/min H_2 gas was used as carrier gas, the precursor gas flow rate of SiC matrix was 100–500 L/min, and the single deposition time was 100–150 h. When the density of the material was more than or equal to 2.5 g/cm^3, the densification process was completed. The surface of the SiC/SiC guide vane was then deposited the SiC coating using the CVD method at approximately 1000 °C. Figure 3.1 shows the SiC/SiC high-pressure turbine single guide vane. Figure 3.2 shows the photograph of the SiC/SiC high-pressure turbine twin guide vanes observed under X-ray Computed Tomography (XCT).

Fig. 3.1 SiC/SiC high-pressure turbine single guide vane

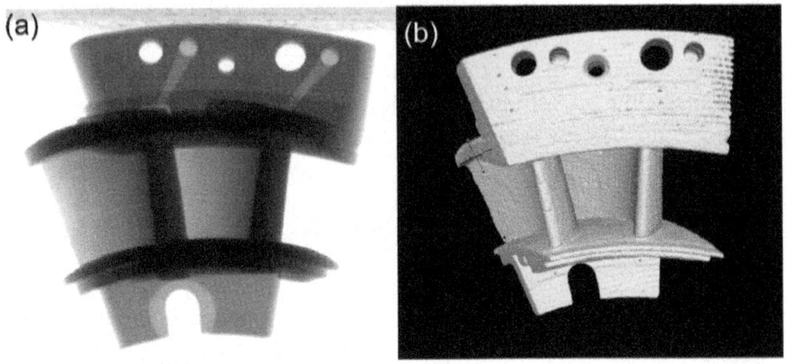

Fig. 3.2 The photograph of SiC/SiC high-pressure turbine twin guide vanes observed under XCT. **a** Front view; and **b** Back view

3.4 Thermal Shock Test Procedures

Thermal shock experiments of CVI SiC/SiC high-pressure turbine guide vanes were carried out on the third-generation turbine service environment simulation device, developed by Xiangtan University, which can simulate single or any two or three kinds of service environment, e.g., alternating temperature cycle, impact of hard particles, erosion of corrosive gas, etc. The schematic of the function principle for the service environment simulator device is shown in Fig. 3.3. The mixture of kerosene and oxygen in the flame gun was ignited using the high-voltage electric spark. Alumina particles were selected as erosion particles, the erosion particle rate can be adjustable from 2.0 to 250 g/min, and the diameter of erosion particles was in the range of 10–300 μm. The particle or flame flow rate is adjustable from 20 to 600 m/s, and the maximum experimental temperature of the sample can reach more than 1700 °C.

The SiC/SiC turbine guide vanes were placed in the high temperature zone and kept for a specified duration, and then transferred to the low temperature zone for cooling. The SiC/SiC guide vanes are fixed on the fixture matching the radian of the

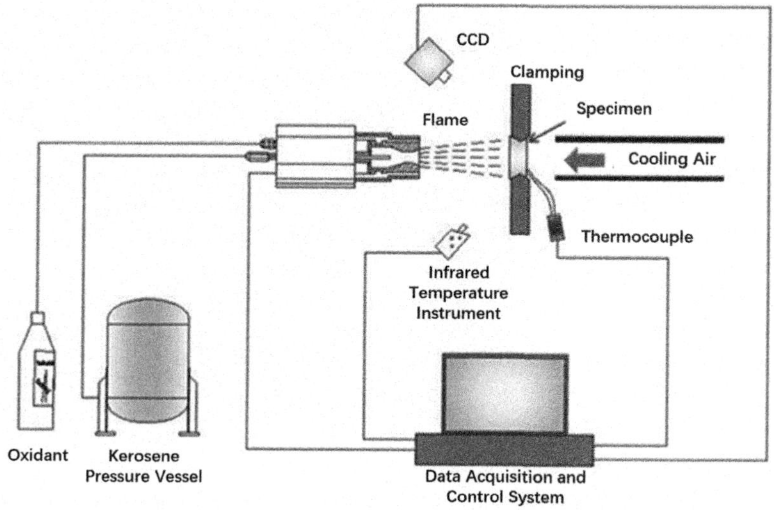

Fig. 3.3 Schematic of function principle for the service environment simulation device

bottom guide vane. The bolt passes through the guide vane edge and is fixed with the fixture. There is a hollow hole from the bottom to the top of the fixture to guide the cooling air into the vane, and the cooling air flows out from the cooling channel inside the vane to provide the cooling air inside the guide vane.

The infrared thermometer MR1SBSF (Raytak, USA) was used to monitor the surface temperature of guide vanes and record the real-time temperature at the guide vane's surface and the flame.

For the single guide vane, the temperature of leading-edge middle section of the guide vane is 1350 °C, heating time is 20 s, holding time is 30 s, cooling time is 50 s. The inlet pressure of the cooling channel of the guide vane is 0.7–0.79 MPa, and the inlet flow rate of the cooling channel is 60 L/min or 77 g/min, and the temperature for the cooling air is 25 °C.

For the twin guide vane, the temperature at the guide vane's leading edge middle section is set to be $T = 1400$, 1450 and 1480 °C with the heating time $t = 30$ s, holding time $t = 30$ s, and the cooling time $t = 60$ s. The inlet pressure of the oxygen in the flame gun was 1.24 MPa, and the flow rate of the oxygen in the flame gun was 180, 200, and 230 L/min for the target temperature of 1400, 1450 and 1480 °C. The inlet pressure of the kerosene in the flame gun was 0.6 MPa, and the flow rate of the kerosene was 5, 6 and 7 L/min the target temperature of 1400, 1450 and 1480 °C. For the testing temperature at $T = 1480$ °C, another thermal shock test with the heating time $t = 30$ s, holding time $t = 60$ s and the cooling time $t = 90$ s is also conducted. The inlet pressure of the cooling channel of the guide vane is $P = 0.7$–0.79 MPa, and the inlet flow rate of the cooling channel is $\rho = 60$ L/min or 77 g/min, and the temperature for the cooling air is $T = 25$ °C.

Fig. 3.4 Thermal shock experiment for the SiC/SiC turbine twin guide vanes

Figure 3.4 shows the figure of the thermal shock test for the turbine guide vane. After cyclic thermal shock tests, the damage regions are observed under scanning electronic microscope (SEM).

3.5 Thermal Shock Test of High-Pressure Turbine Single Guide Vane

3.5.1 Surface Temperature Distributions

Under the thermal shock test, the surface temperature distribution for the single guide vane is shown in Fig. 3.5. The temperature in the middle of the leading edge was the highest, i.e., 1350 °C. The temperature of the basin area decreased gradually from the leading edge to the trailing edge, and the temperature of the back of the vane was the lowest. The overall temperature of the basin area was much higher than that of the back area of the vane. When the internal region of the vane was flown with cooling air, the temperature of the five measuring points of the vane was lower than that without cooling air, and the overall temperature change trend was consistent with that without cooling air.

By extracting the infrared temperature sequence data of the five measuring points, five temperature curves were shown in Fig. 3.6. It can be obtained from the figure that the leading-edge temperature is the highest, the temperature of the basin area of the vane is the second, the temperature of the back of the vane is the lowest, which is consistent with the cloud chart of the temperature distribution of infrared thermal imaging. The leading-edge position of the vane can be heated to 1350 °C of the target setting temperature within 20 s and can be well insulated for 30 s. The temperature value in the insulation stage is relatively constant and the fluctuation is very small. After cooling, the temperature value is about 370 °C, and the temperature

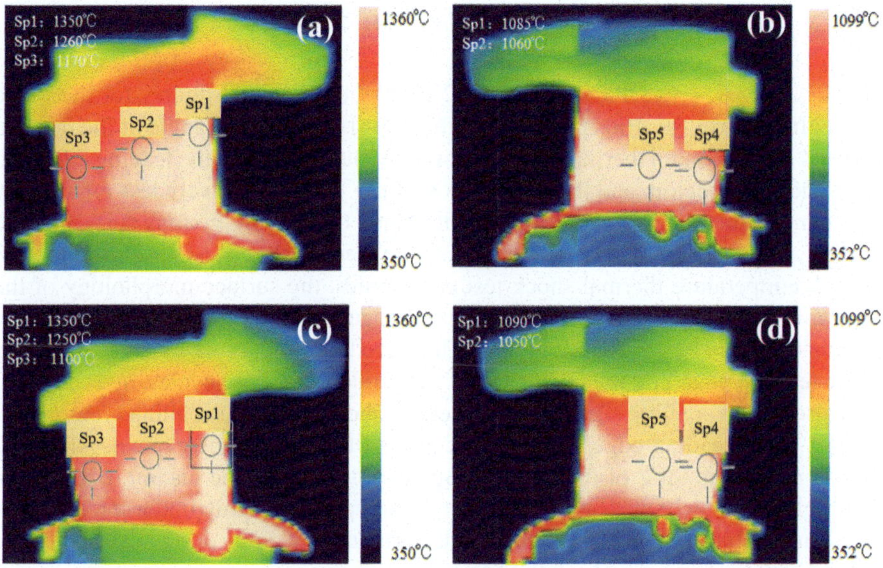

Fig. 3.5 Surface temperature distribution for the single guide vane under thermal shock test **a** basin area without cooling air; **b** back area without cooling air; **c** basin area with cooling air; and **d** back area with cooling area

of other measuring points is lower than the front edge temperature. The maximum temperature at the tail edge can reach the required 1077 °C, and the temperature fluctuation of each measuring point is less than ±10 °C.

Fig. 3.6 The temperature change with time during thermal shock test for five different measurement spots

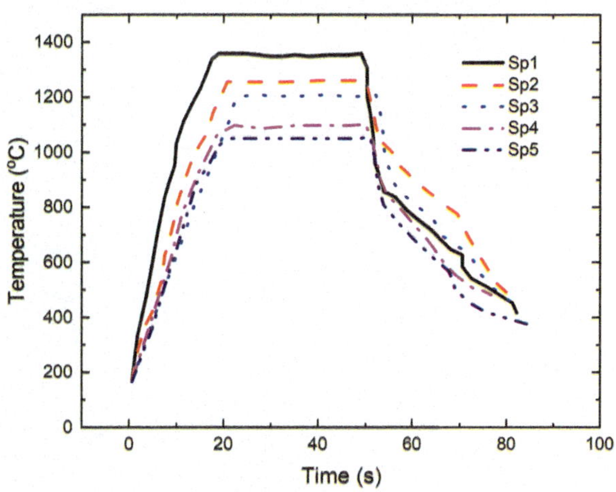

3.5.2 Damage Analysis After Thermal Shock

Figures 3.7, 3.8, 3.9, 3.10, 3.11 and 3.12 show the schematic figures for the guide vanes after 40, 60, 80, 100, 140 and 160 thermal shocks. As the angle of the thermal shock is 90°, the main thermal shock position of the flame is in the leading edge and the blade basin region, and the back region of the vane is relatively less affected by high temperature thermal shock. The results show that the SiC coating surface of the turbine guide vane is smooth and tidy before the thermal shock test. With the increase of high temperature thermal shock test cycle times, the surface morphology of the back area of the guide vane has no obvious change, but the coating peeling off in the leading edge and basin area is obvious. There are large spalling areas of SiC coating in the leading-edge and the basin area of the guide vane at 40 cycles, and the basin area also shows spalling signs after 60 thermal shocks, and the spalling areas of SiC coating at the leading edge and the back area of the guide vane expand. After 160 cycles, more than 90% of the SiC coating spalling area on the leading-edge, basin and back area is achieved, and the whole guide vane turns gray under the action of high temperature gas, which is due to the formation of SiO_2 on the surface of the SiC/SiC guide vane under high temperature environment.

Figure 3.13 shows the damage observation of the SiC coating of SiC/SiC guide vane after thermal shock. It can be found that the SiC coating peeling off occurs on the surface of the SiC/SiC guide vane. In the spalling area, the white region is distributed around the surface due to the oxidation of SiC coating at high temperature.

Fig. 3.7 Schematic of guide vane after 40 thermal shock tests **a** leading-edge; **b** basin area; and **c** back area

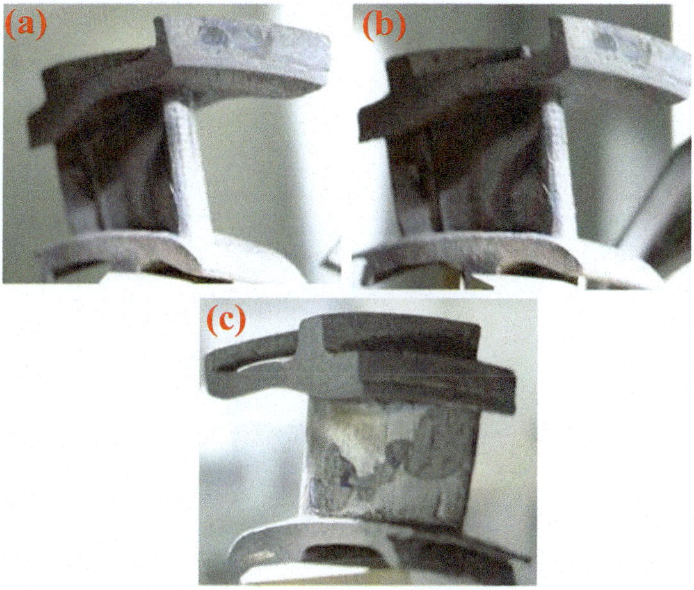

Fig. 3.8 Schematic of guide vane after 60 thermal shock tests **a** leading-edge; **b** basin area; and **c** back area

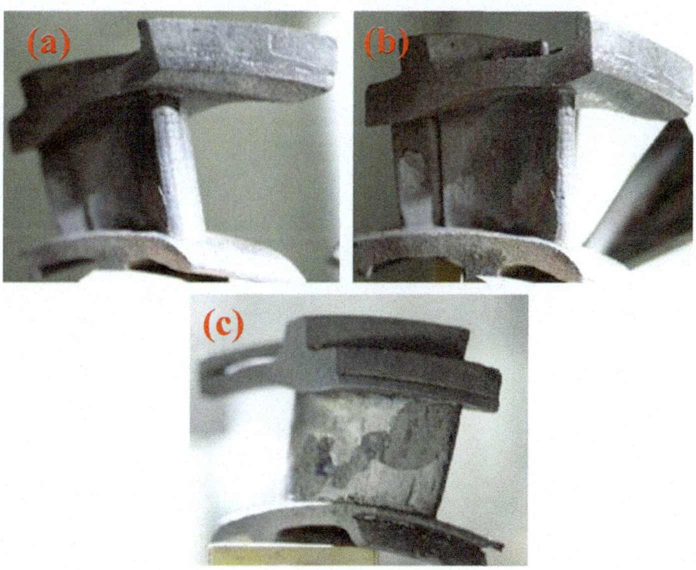

Fig. 3.9 Schematic of guide vane after 80 thermal shock tests **a** leading-edge; **b** basin area; and **c** back area

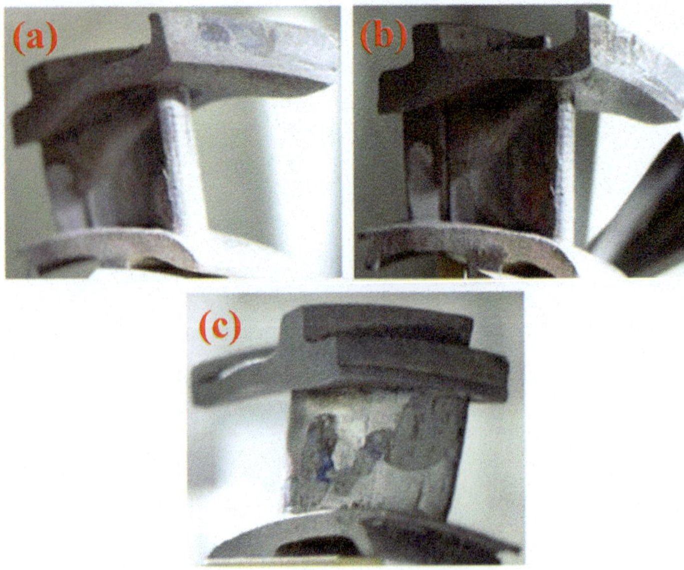

Fig. 3.10 Schematic of guide vane after 100 thermal shock tests **a** leading-edge; **b** basin area; and **c** back area

Fig. 3.11 Schematic of guide vane after 140 thermal shock tests **a** leading-edge; **b** basin area; and **c** back area

Fig. 3.12 Schematic of guide vane after 160 thermal shock tests **a** leading-edge; **b** basin area; and **c** back area

Figure 3.14 shows the oxidation distribution of the SiC coating on the surface of the SiC/SiC guide vane after thermal shock tests. The formation of SiO_2 on the surface of SiC coating after oxidation caused the degradation of the bonding strength between the coating and the SiC/SiC composite leading to the peeling off of the SiC coating.

3.6 Thermal Shock Test of High-Pressure Turbine Twin Guide Vane

3.6.1 Surface Temperature Distributions

Under thermal shock test, the infrared thermal imaging of twin guide vanes at the target temperature of $T = 1400$ °C is shown in Fig. 3.15. Due to the shielding of the guide vane, only part of the temperature field of a single guide vane can be monitored and measured (i.e., the basin area in the left guide vane and the back area in the right guide vane). For the left guide vane, the temperature in the middle of the leading edge was the highest, i.e., $T = 1412$ °C. The temperature at the basin area decreased gradually from the leading edge to the trailing edge, and the temperature of the trailing edge was the lowest. For the right guide vane, the temperature of the trailing edge and leading edge was the highest, and the temperature of the middle

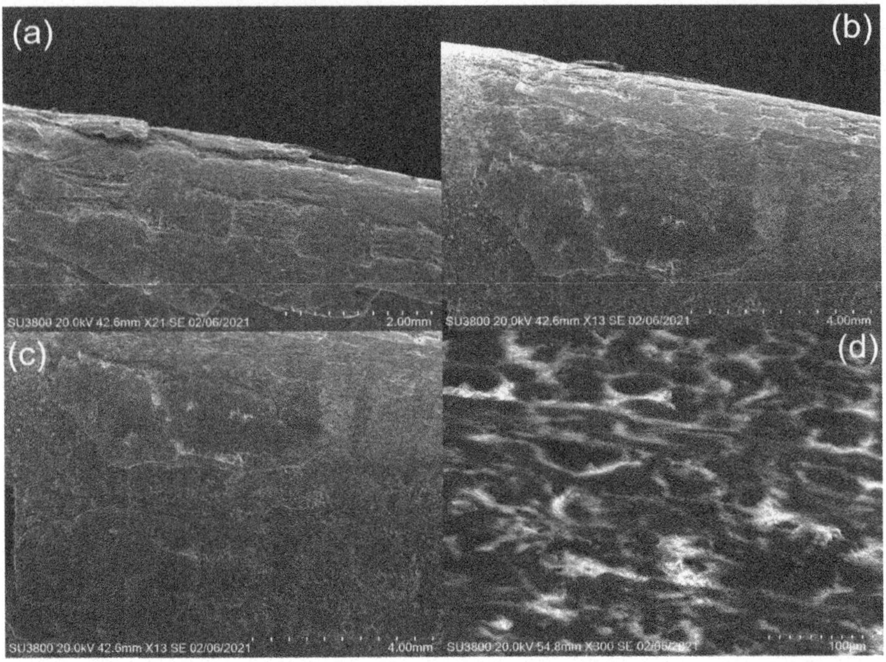

Fig. 3.13 Damage observation of the SiC coating of SiC/SiC guide vane after thermal shock test

region was the lowest. The temperature at the basin area was higher than that at the back area of the guide vane.

Figure 3.16 shows the extracting infrared temperature data of six measuring points for the left and right guide vanes. It can be found that the temperature at the leading edge of the left guide vane is the highest, and the temperature at the back area of the right guide vane is the lowest, which is consistent with the cloud chart of the temperature distribution of infrared thermal imaging. The measurement positions of the left guide vane can be heated to 1400 °C of the target setting temperature within 30 s and can be well held for 30 s. The temperature value in the holding stage is relatively constant and the fluctuation is small. The heating rate at the back area of the right guide vane is much slower than that at the basin area. After cooling, the temperature value is about 630 °C, and the temperature of other measuring points is lower than the front edge temperature. The maximum temperature at the leading edge of the left guide vane can reach the required 1400 °C, and the temperature fluctuation of each measuring point during the holding stage is less than ±20 °C.

Figure 3.17 shows the infrared thermal imaging of twin guide vanes under thermal shock test at the target temperature $T = 1480$ °C. Due to the shielding of the guide vane, only part of the temperature field of a single guide vane can be monitored and measured (i.e., the basin and back area in the left guide vane). The temperature in the middle of the leading edge was the highest, i.e., $T = 1479$ °C. The temperature at the basin area decreased gradually from the leading edge to the trailing edge, and

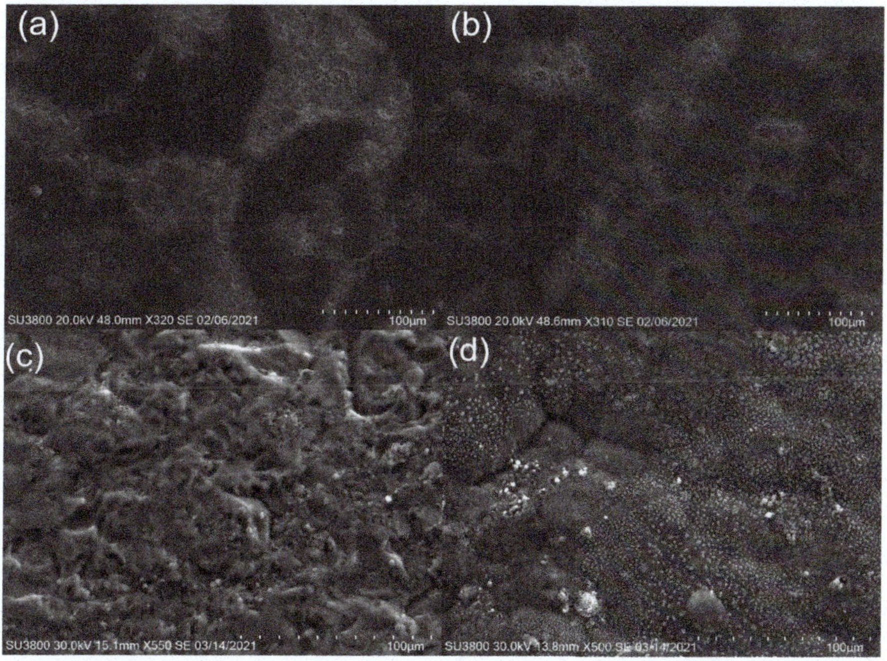

Fig. 3.14 Oxidation of SiC coating on SiC/SiC guide vane after thermal shock

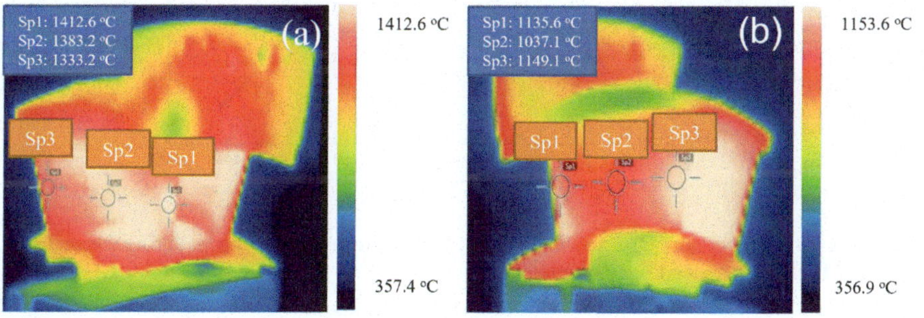

Fig. 3.15 Surface temperature distribution for the twin guide vanes under thermal shock test at the target temperature of $T = 1400$ °C. **a** Basin area of the left guide vane; and **b** Back area of the right guide vane

the temperature of the trailing edge was the lowest. For the back area of the left guide vane, the temperature of the trailing edge was low, and the middle region was high, and the temperature at the basin area was higher than that at the back area of the guide vane.

Figure 3.18 shows the extracting infrared temperature data of five measuring points for the left guide vane. It can be found that the temperature at the leading

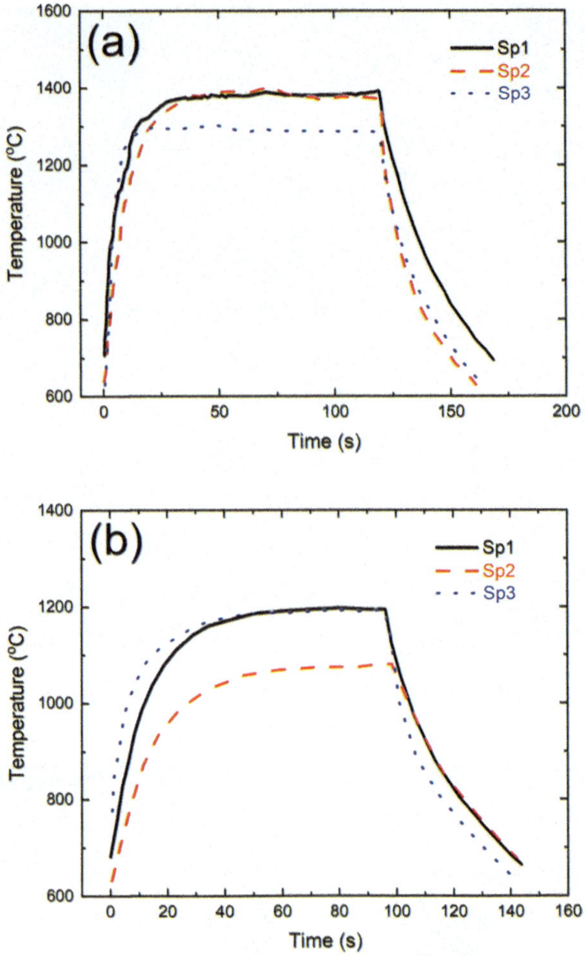

Fig. 3.16 The temperature change with time for different measurement spots during thermal shock test. **a** Left guide vane; and **b** Right guide vane

edge of the left guide vane is the highest, and the temperature at the back area of the left guide vane is the lowest, which is consistent with the cloud chart of the temperature distribution of infrared thermal imaging. The measurement positions of the left guide vane can be heated to 1480 °C of the target setting temperature within 30 s and can be well held for 30 s. The temperature value in the holding stage is relatively constant with small fluctuation. The heating rate at the back area of the left guide vane is much slower than that at the basin area. After cooling, the temperature value is approximately 600 °C, and the temperature of other measuring points is lower than the front edge temperature. The temperature fluctuation of each measuring point during the holding stage is less than ±20 °C.

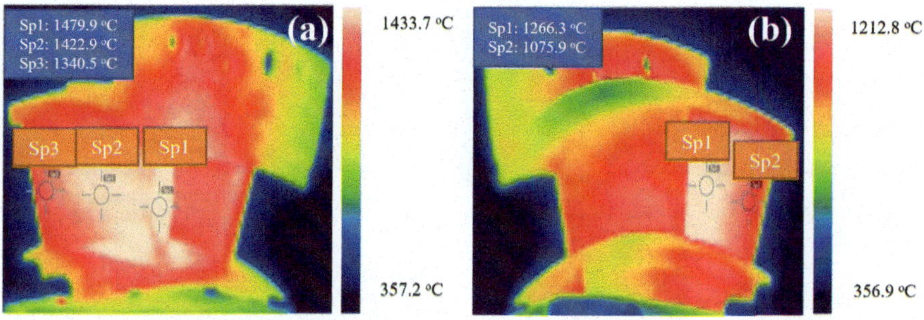

Fig. 3.17 Surface temperature distribution for the twin guide vanes under thermal shock test at the target temperature of $T = 1480$ °C. **a** Basin area of the left guide vane; and **b** Back area of the right guide vane

Fig. 3.18 The temperature change with time for five different measurement spots during thermal shock test of left guide vane

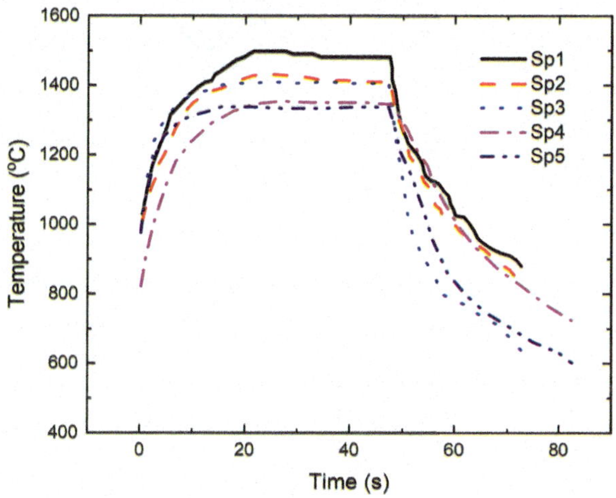

3.6.2 Damage Analysis After Thermal Shock

Figure 3.19 shows the schematic figures for the twin guide vanes at leading edge, basin and back region after 100, 200, 300 and 400 thermal shock tests. Because of the angle of the thermal shock is 90°, the main thermal shock position of the flame is in the leading edge and the basin region, and the back region of the guide vane is relatively less affected by the high temperature thermal shock.

Before the thermal shock test at $T = 1400$ °C, the surface of guide vane with the SiC seal coating is smooth. With the increase applied cycle number of high temperature thermal shock test, the surface morphology of the back area at the guide vane has no obvious change, however, the coating peeling off in the leading edge at the basin area becomes obvious, which is the peeling off the surface sealing SiC coating of the SiC/SiC guide vane. At $N = 100$ cycles, large spalling areas appeared on the leading edge, the basin and the back region of the guide vane. After $N = 400$

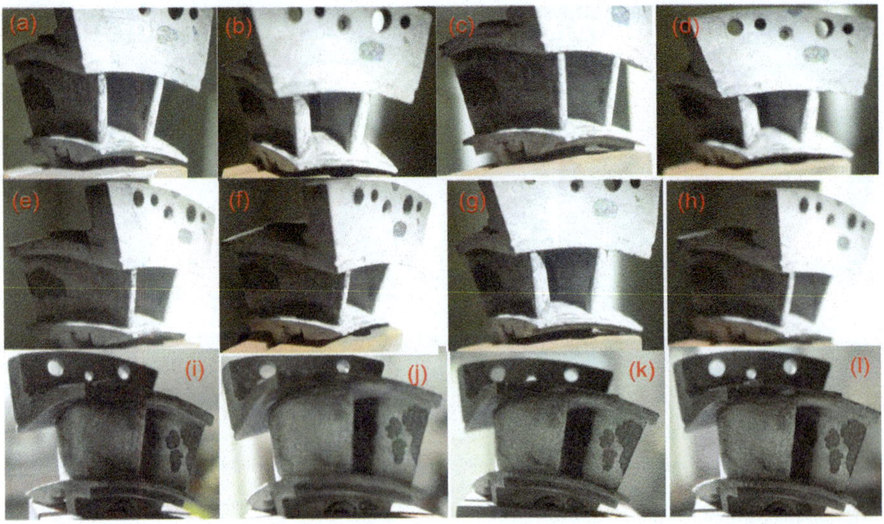

Fig. 3.19 Thermal shock damage of SiC/SiC twin guide vane at target temperature of $T = 1400$ °C for leading edge at **a** $N = 100$, **b** $N = 200$, **c** $N = 300$, and **d** $N = 400$, the basin region at **e** $N = 100$, **f** $N = 200$, **g** $N = 300$, and **h** $N = 400$, and the back region at **i** $N = 100$, **j** $N = 200$, **k** $N = 300$, and **l** $N = 400$

cycles, the peeling area of the coating at the basin and back area reaches more than 30%, and the guide vane turns gray under the action of high temperature gas. This is due to the formation of SiO_2 on the surface of SiC/SiC guide vane due to high temperature oxidation under high temperature environment (Li 2021b, 2017).

When the thermal shock testing temperature increased to $T = 1450$ °C, with increasing thermal shock cycle number, the spalling area at the basin and back region of the guide vane was not affected by the increased temperature. Figure 3.20 shows the figures for the left guide vane at leading edge, basin region, and back region after 100, 200, and 300 thermal shock tests. When the thermal shock cycle number increased to $N = 300$, delamination occurred at the tenon of the left guide vane, as shown in Fig. 3.21. Under thermal shock test, the temperature difference would cause microcracking in the SiC ceramic matrix, and the propagation of these microcracks may cause delamination in the components.

Increasing the thermal shock testing temperature to $T = 1480$ °C and increasing the hold time from $t = 30$ to 60 s, the spalling area at the basin and back region did not increase significantly, as shown in Figs. 3.22 and 3.23. However, at the thermal shock cycle number of $N = 200$, new delamination occurred at the trailing edge of the guide vane, as shown in Fig. 3.24.

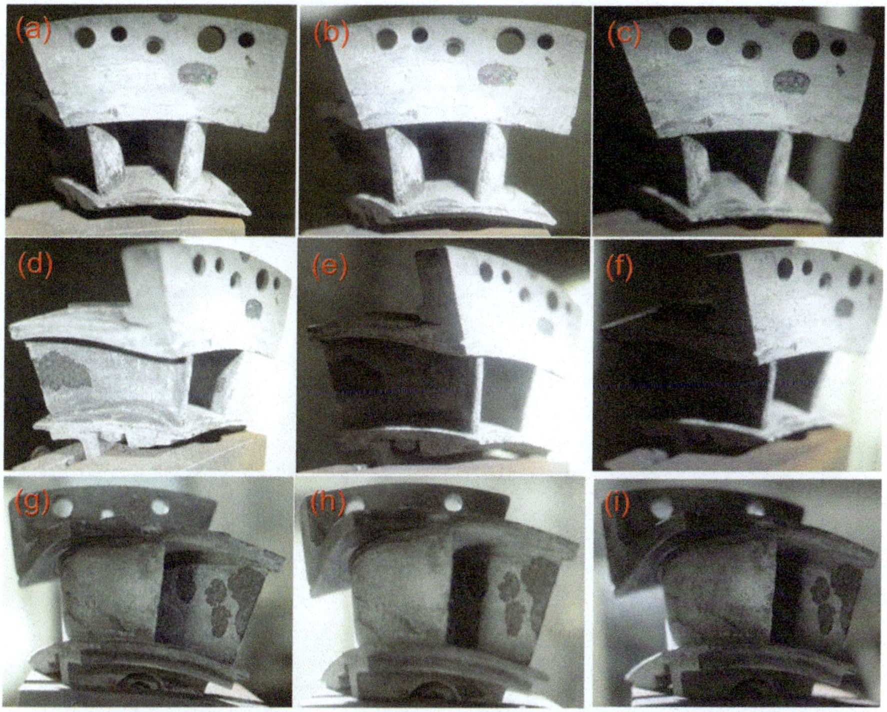

Fig. 3.20 Thermal shock damage of SiC/SiC twin guide vane at target temperature of $T = 1450$ °C for leading edge at **a** $N = 100$, **b** $N = 200$, and **c** $N = 300$, the basin region at **d** $N = 100$, **e** $N = 200$, and **f** $N = 300$, and the back region at **g** $N = 100$, **h** $N = 200$, and **i** $N = 300$

Fig. 3.21 Thermal shock test of SiC/SiC twin guide vane at target temperature of $T = 1450$ °C at $N = 300$

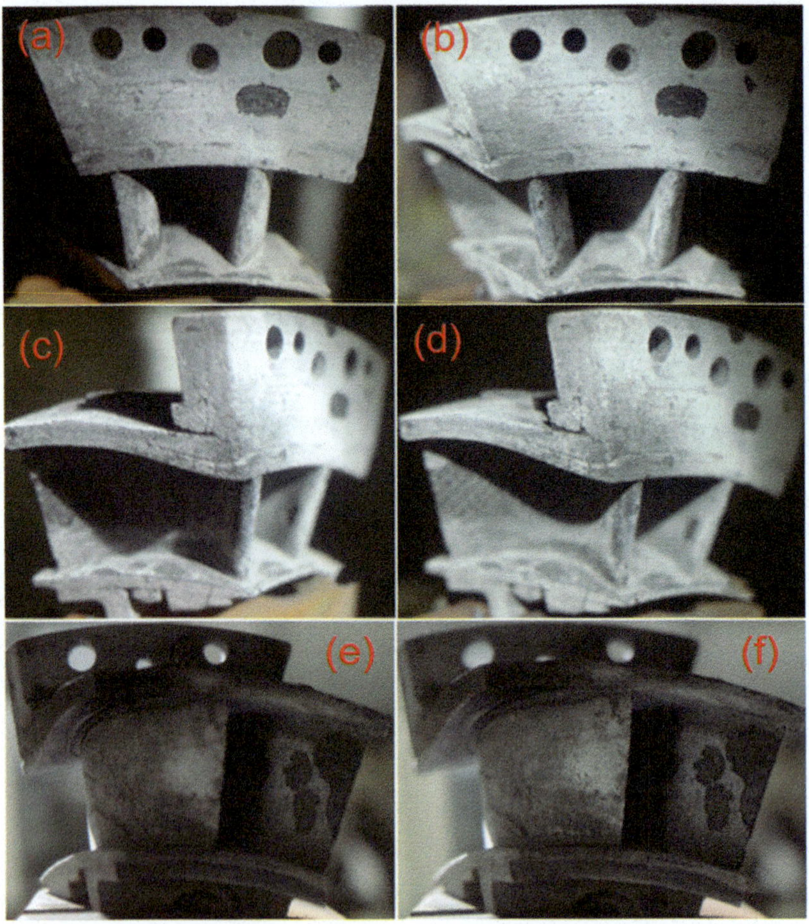

Fig. 3.22 Thermal shock damage of SiC/SiC twin guide vane at target temperature of $T = 1480$ °C with the hold time of $t = 30$ s for leading edge at **a** $N = 100$, and **b** $N = 200$, the basin region at **c** $N = 100$, and **d** $N = 200$, and the back region at **e** $N = 100$, and **f** $N = 200$

3.7 Summary and Conclusions

In this chapter, the SiC/SiC high-pressure single/twin turbine guide vanes were designed and fabricated using the CVI method. The thermal shock tests were conducted to investigate the thermal shock damage evolution. The surface temperature field of the SiC/SiC guide vane was analyzed and the damage characteristics of the composite guide vane after experiencing different thermal shocks were discussed.

(1) For the single turbine guide vane, the temperature in the middle of the leading edge is the highest, i.e., 1350 °C. The temperature of the basin area decreased gradually from the leading edge to the trailing edge, and the temperature of

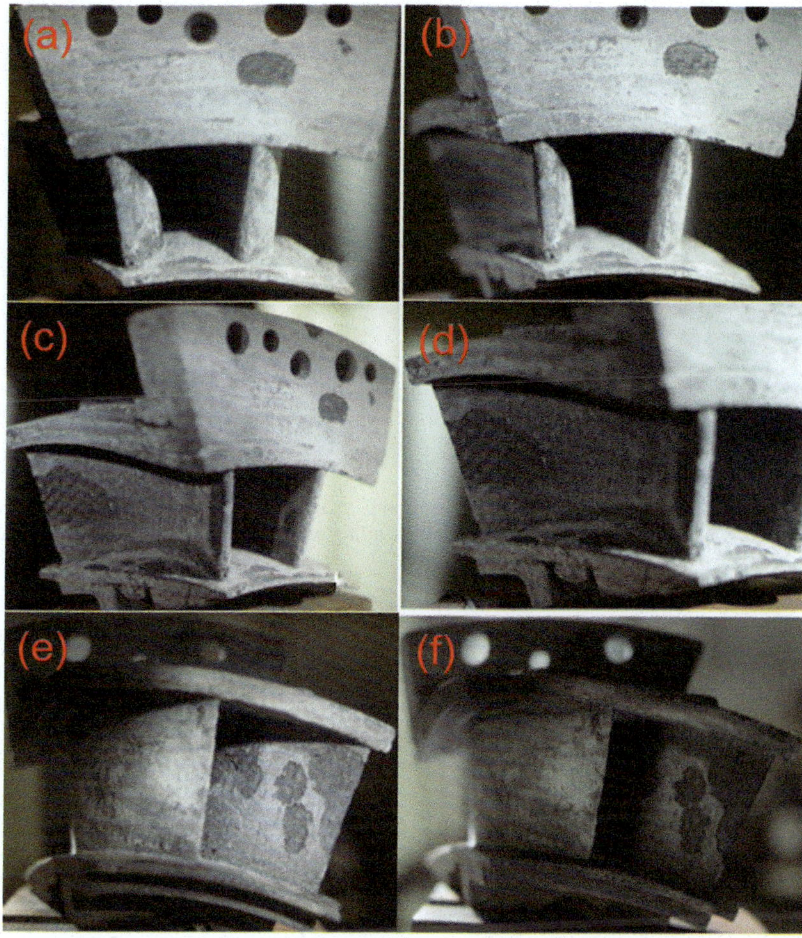

Fig. 3.23 Thermal shock damage of SiC/SiC twin guide vane at target temperature of $T = 1480$ °C with the hold time of $t = 60$ s for leading edge at **a** $N = 100$, and **b** $N = 200$, the basin region at **c** $N = 100$, and **d** $N = 200$, and the back region at **e** $N = 100$, and **f** $N = 200$

the back of the vane was the lowest. The overall temperature of the basin area is much higher than that of the back area of the vane. After 160 cycles, more than 90% of the SiC coating spalling area on the leading-edge, basin and back area is achieved, and the whole guide vane turns gray under the action of high temperature gas, which is due to the formation of SiO_2 on the surface of SiC/SiC guide vane under high temperature environment.

(2) For the twin turbine guide vane, before the thermal shock test, the surface of the turbine guide vane is smooth. After 100 cycles, large spalling areas appeared on the leading edge and back region, which were the spalling of SiC sealing coating on the surface of the SiC/SiC guide vane. After 400 cycles, the spalling area of

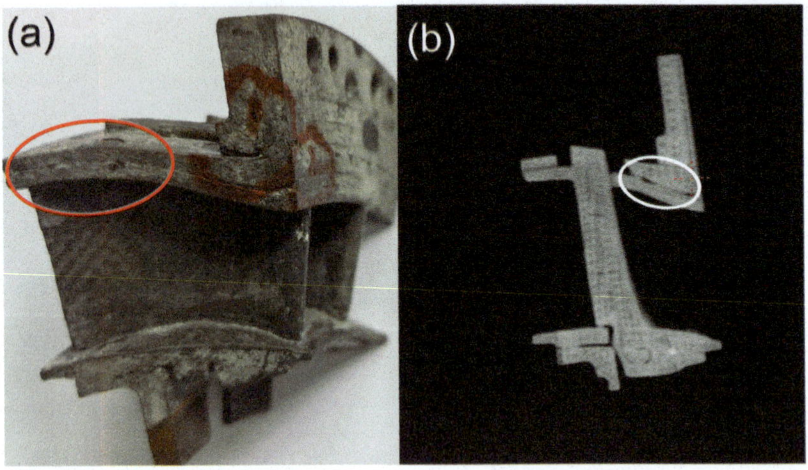

Fig. 3.24 Thermal shock test of SiC/SiC twin guide vane at target temperature of $T = 1480$ °C at $N = 200$. **a** New delamination occurred at the trailing edge; and **b** New delamination observed under XCT

the coating at the basin and back region of the guide vane is more than 30%, and the whole guide vane becomes gray under the action of high temperature gas, due to the formation of SiO_2 on the surface of SiC/SiC guide vane under high temperature environment. When the thermal shock test temperature was raised to 1450 °C, the spalling area of the basin and the back region of the guide vane did not increase significantly, but the delamination occurred at the tenon of the guide vane. When the thermal shock test temperature was increased to 1480 °C and the holding time increased from 30–60 s, the spalling area of the basin and back region of the guide vane did not increase significantly, but a new delamination phenomenon appeared on the upper surface of the guide vane near the trailing edge.

References

Fumiaki W, Takeshi N, Yousuke M (2017) Design and testing for ceramic matrix composite turbine vane. In: Proceedings of ASME turbo expo 2017: turbomachinery technical conference and exposition GT2017 June 26–30, 2017, Charlotte, NC, USA

Li L (2017) Effects of temperature and oxidation on cyclic-fatigue life of 2D woven ceramic-matrix composites. J Aerosp Eng 30(4):0401700

Li L (2021a) High temperature mechanical behavior of ceramic-matrix composites. Wiley-VCH, Weinheim, Germany

Li L (2021b) Modeling stress-dependent matrix multiple fracture of fiber-reinforced ceramic-matrix composites considering fiber oxidation and fracture. Compos Interfaces 28(4):329–361

Li L (2022) High-temperature mechanical hysteresis behavior of ceramic-matrix composites. CRC Press, Taylor & Francis Group, London, UK

Li L (2023) Ceramic matrix composites: lifetime and strength prediction under static and stochastic loading. Elsevier, Oxford, UK

Li L (2024) Micromechanics of ceramic-matrix composites at elevated temperatures. Springer Nature, Singapore

Liu X, Guo X, Xu Y, Li L, Zhu W et al (2021) Cyclic thermal shock damage behavior in CVI SiC/SiC high-pressure turbine twin guide vanes. Materials 14:6104

Verrilli M, Calomino A, Craig Robinson R (2004) Ceramic matrix composite vane subelement testing in a gas turbine environment. In: Proceedings of ASME turbo expo 2004, power for land, sea, and air, June 14–17, 2004, Vienna, Austria

Chapter 4
Design, Fabrication and Testing of CMC Combustor Liner

Abstract In this chapter, the design, fabrication and testing of CMC combustor liner were investigated. The service performance analysis of the combustor liner was at first analyzed. The SiC/SiC combustor liner with the dimension of 296.48 mm (maximum diameter of outer liner) × 149.85 mm (minimum diameter of inner liner) × 100.46 mm (height), a wall thickness of 3–10 mm (including splicing structure) was fabricated using the CVI process. The surface of the SiC/SiC combustor liner was coated with EBC (Environmental Barrier Coatings, EBC) to improve the reliability and durability. Pulsed laser micromachining technology was used to realize the precise processing of SiC/SiC combustor liner micro-holes, and ultrasound-assisted machining technology was used to meet the demand for high-precision profile processing for various air-film holes with diameters larger than 0.5 mm, and the roughness can reach Ra = 0.3~0.5. After the inner and outer liners and other components have been densified and processed, the SiC rivets are used for on-line riveting, which achieves a connection result equivalent to "metal riveting" after CVI-SiC deposition, improves the ability of components to maintain the profile, and prevents delamination of combustor liner during the service life. On October 14, 2022, the ground ignition and performance test of the combustor liner was conducted. The average gas temperature was 1426 °C, with localized hot spots reaching 1734 °C. Localized EBC coating peeling was found after the ground testing.

Keywords SiC/SiC · Combustor liner · Environmental barrier coatings · Ground testing

4.1 Introduction

The combustor liner is an important component of the turbine engine, whose main function is to convert the heat energy of the high-temperature and high-pressure airflow behind the combustion chamber into kinetic energy, and ensure that the gas flows out in a certain direction, pushing the turbine rotor to do work, and ensuring that the engine can output stable and continuous power. The high temperature resistance

X. Liu and L. Li, *Design, Fabrication and Testing of Aeroengine Ceramic-Matrix Composite Components*, Advanced Ceramics and Composites 7,
https://doi.org/10.1007/978-981-97-6109-8_4

of existing metal materials can no longer meet this requirement, and ceramic-matrix composites (CMCs) are a new type of integrated thermal structure/function material, and possess low density, high temperature resistance, high specific strength, high specific modulus, oxidation resistance, ablation resistance, insensitivity to cracks, and no catastrophic damage (Li 2023a, 2022, 2021a, 2021b). There are generally two types of CMCs for combustor liner: one is the silicon carbide fiber reinforced silicon carbide matrix composite, i.e., SiC/SiC composite; the other type is the oxide fiber reinforced oxide matrix composite, i.e., oxiede/oxide composite materials (Li 2021c, 2023b).

GE has manufactured and validated a SiC/SiC low emission combustor liner with multiple inclined cooling hole (Padture 2016), as shown in Fig. 4.1. Experiments have shown that using this SiC/SiC combustor liner can reduce the cooling air of the combustor by 50%, reduce the mass by 50%, and reduce the NO_x emissions by about 20%. The GE9X engine expands the CMC use to the inner and outer combustor liners, High-Pressure Turbine (HPT) stage 1 and stage 2 nozzles and stage 1 shrouds. The inner and outer liners of the combustor of the GE9X engine were made of SiC/SiC composites, which significantly increased the liner's durability, reduce the cooling air usage and NO_x emissions by 25–30%, raised the combustor exit temperature by 150 K, exhibited excellent high-temperature performance, and significantly enhanced the overall performance of the aero-engine.

UTRC (United Technologies Research Center) and P&W Canada validated the SiC/SiC combustor in a PW 200 series combustor (Bhatia et al. 2010), as shown in Fig. 4.2. The CMC combustor offers the potential elimination of the film cooling to provide reduced emissions and increased durability. The high temperature capability of the CMC is expected to increase the life of the combustor components while enabling the elimination of film cooling and a reduction of combustor cooling. The thermal analysis showed that the maximum temperatures on the CMC are well within the average capability for melt infiltration (MI) SiC/SiC composite, and the stress analysis also found that the stresses in PW 206 components was within the SiC/SiC material's average capability. The fill annular CMC combustor rig was engine

Fig. 4.1 Low emission SiC/SiC combustor liner developed by GE company

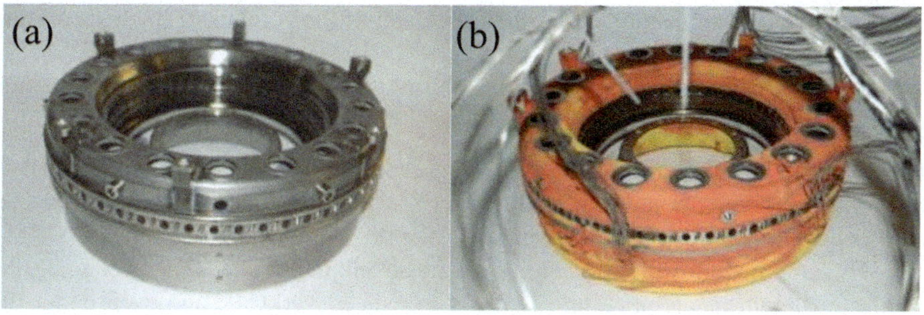

Fig. 4.2 MI SiC/SiC combustor **a** before and **b** after rig test

tested for 250 cycles between idle (40,000 rpm) and full power (57,000 rpm) and severely tested the response of the CMC/metal interfaces to accelerated thermal cycling. The test was stopped after 250 cycles with no damage observed. The use of CMC combustor in P&W200 series demonstrated a 40–50% improvement in pattern factor, a 30% reduction in NO_x at full power and a 20% reduction in CO at idle condition compared to the baseline metal combustor.

In this chapter, the design, fabrication and testing of CMC combustor liner were investigated. The service performance analysis of the combustor liner was at first analyzed. The SiC/SiC combustor liner was fabricated using the CVI process, and coated with Environmental Barrier Coatings (EBC) to improve the reliability and durability. Pulsed laser micromachining technology was used to realize the precise processing of SiC/SiC combustor liner micro-holes, and ultrasound-assisted machining technology was used to meet the demand for high-precision profile processing for various air-film holes with diameters larger than 0.5 mm. After the inner and outer liners and other components were densified and processed, the SiC rivets were used for on-line riveting. On October 14, 2022, the ground ignition and performance test of the combustor liner was conducted.

4.2 Service Performance Analysis of Combustor Liner

The working environment of the aero-engine combustion chamber is extremely harsh (Lacombe et al. 2009; Misra 2001). The combustion process is carried out in a high-speed airflow and a lean oil mixture. The structural components of the combustion chamber work under high temperature and high stress, bearing static and vibration loads generated by aerodynamic and inertial forces, as well as being corroded by corrosive gas atmospheres. After the engine starts, the temperature of the combustion gas is approximately 1800–2000 °C, far exceeding the tolerance temperature of the existing combustion chamber and turbine structure. Therefore, unburned air is usually gradually introduced into the combustor liner, partly to reduce the temperature of the gas and partly to cool the walls of the combustor liner. Film cooling technology is usually used to allow air to flow along the wall of the combustor liner, separating it from hot gas.

The main environmental conditions during the operation of the combustion chamber (mainly the combustor liner) of an aircraft engine include:

(1) Poor fuel gas environment: containing excessive oxygen, nitrogen, as well as water vapor, carbon dioxide, nitrogen oxides, and sulfides generated after fuel combustion, with a gas flow velocity of approximately 50–100 m/s.
(2) High temperature environment: With the development of advanced engines, the mainstream gas temperature is likely to exceed 1900 °C, and even under the condition of gas film cooling, the wall temperature of the combustor liner may reach 1300 °C or even higher.
(3) Temperature gradient environment: The inner and outer walls of the combustor liner are respectively subjected to gas heating and air cooling. The temperature difference on both sides of the wall is about 250–300 K. Based on an estimated wall thickness of 2 mm, the temperature gradient of the combustor liner in the thickness direction can reach 150 K/mm.
(4) Stress environment: mainly caused by assembly stress, inertia force, and vibration stress, stress caused by pressure difference of about 20 kPa on both sides of the combustor liner wall, thermal stress caused by thickness direction and axial temperature gradient of the combustor liner wall, and thermal stress caused by thermal deformation mismatch in high temperature environment due to connection of different materials.

According to the working environment of the combustion chamber in engines, the combustor liner components should meet the following design requirements:

(1) Material selection requirements: able to work for a long time in a high-temperature gas environment without oxidation, erosion, softening or melting; the thermal conductivity should be as high as possible and the coefficient of thermal expansion should be as low as possible; sufficient modulus and strength need to be maintained in both normal and high temperature working environments.
(2) Structural requirements: The components and their connections should have sufficient strength and rigidity, good cooling and reliable thermal compensation to reduce thermal stress; the connection between different materials needs to consider thermal expansion matching design; the structure should be as simple as possible, with good usability and maintenance.

4.3 Design of CMC Combustor Liner

The 3D schematic of the SiC/SiC combustor liner is shown in Fig. 4.3, with dimension of 296.48 mm (maximum diameter of outer liner) × 149.85 mm (minimum diameter of inner liner) × 100.46 mm (height), a wall thickness of 3–10 mm (including splicing structure). The surface of the component includes 720 circular hole structures (including installation holes, cooling holes, etc.), with hole size distribution ranging from 3–26 mm.

The SiC/SiC combustor liner adopts a split design, with a metal adapter ring on the outer edge of the inner liner, as shown in Fig. 4.4. The SiC/SiC inner liner and the metal adapter ring are riveted together using metal rivets; The combustor inner and outer liners are riveted with ceramic rivets, then, the SiC matrix is deposited inside the rivet pores through the CVI process to complete the riveting connection of the inner and outer liners.

The CMC combustor liner uses third-generation SiC fibers as raw materials, and the properties of the SiC fiber are listed in Table 4.1. The preform type is a 2.5D fiber structure, as shown in Fig. 4.5. The weft yarn direction is designed as the axial direction of the combustor liner and the warp yarn direction is designed as the

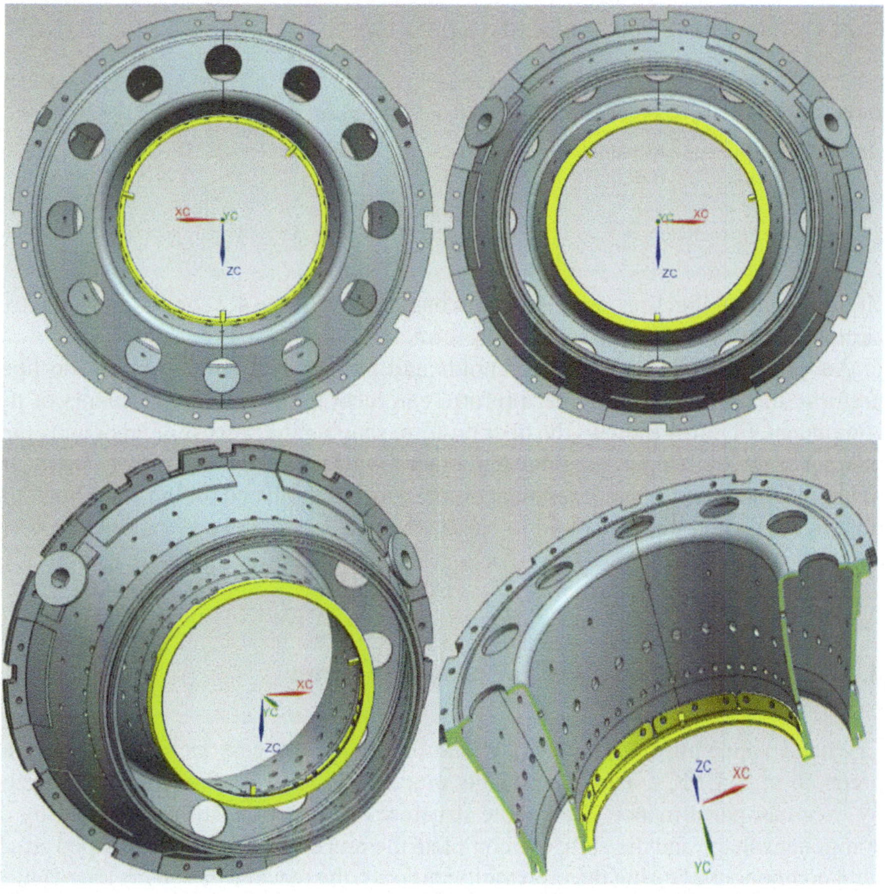

Fig. 4.3 3D schematic of SiC/SiC combustor liner

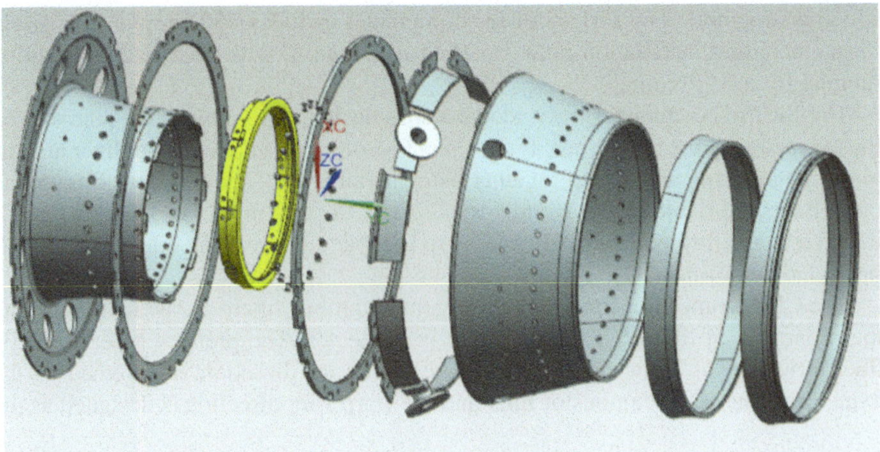

Fig. 4.4 Schematic of SiC/SiC integrated combustor liner

Table 4.1 Material properties of third-generation of SiC fiber

Tensile strength (GPa)	Modulus (GPa)	Density (g/cm^3)	Diameter (μm)	C:Si	O (wt%)
3.22	340	3.10	12.51	1.10	0.58

circumferential direction yarn of the combustor liner. Table 4.2 shows the geometric parameters of the 2.5D woven fiber preform.

Adopting internal and external molds and pre-stressed bonding with the fiber preform, the surface of the fiber preform can meet the design requirements of the component. Figure 4.6 shows the fiber preform shaping the process of inner and outer liners. For the inner liner, the inner and outer molds are used to accurately shape the fiber preform using the pre-stressing compression method; and for the outer liner, the fiber is warped on the core mold to ensure the perfect fit between the fiber preform and core mold.

4.4 Fabrication of CMC Combustor Liner

Chemical vapor infiltration (CVI) is one of the few process technologies for the preparation of CMCs. The components prepared by this process are characterized by excellent performance, designable structure and function, strong adaptability of component shape and the ability to complete the preparation of various shaped structural components. During the preparation process, the reactant gas enters the reaction deposition furnace according to a certain ratio. When the gas flows through the porous preform with the main gas flow, it transfers to the porous preform through the mass

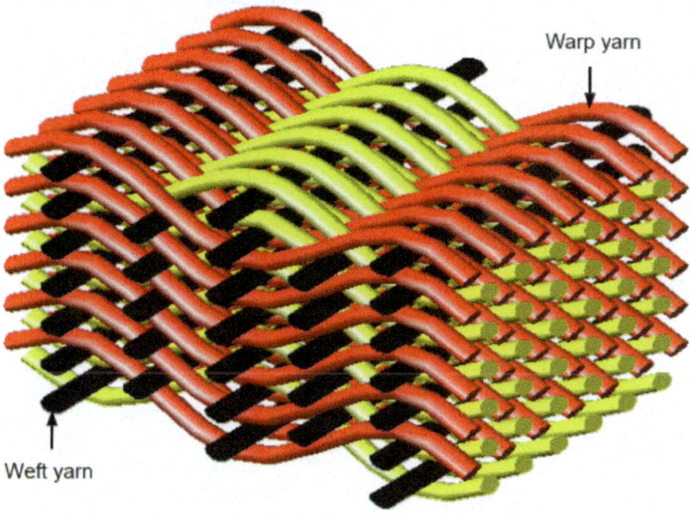

Fig. 4.5 Schematic of 2.5D fiber preform

Table 4.2 Geometric parameters of the 2.5D woven fiber preform

Tows (k)	Warp density (ends/cm)	Number of warp layers	Weft density (picks/cm)	Number of weft layers	Preform thickness (mm)	Crimp angle of warp yarn (°)
6	10	6	3	7	4.2	13

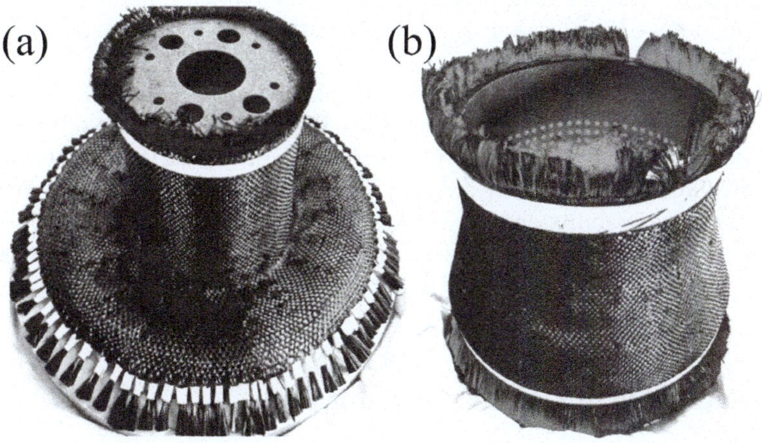

Fig. 4.6 **a** Fiber preform of inner combustor liner; and **b** Fiber preform of outer combustor liner

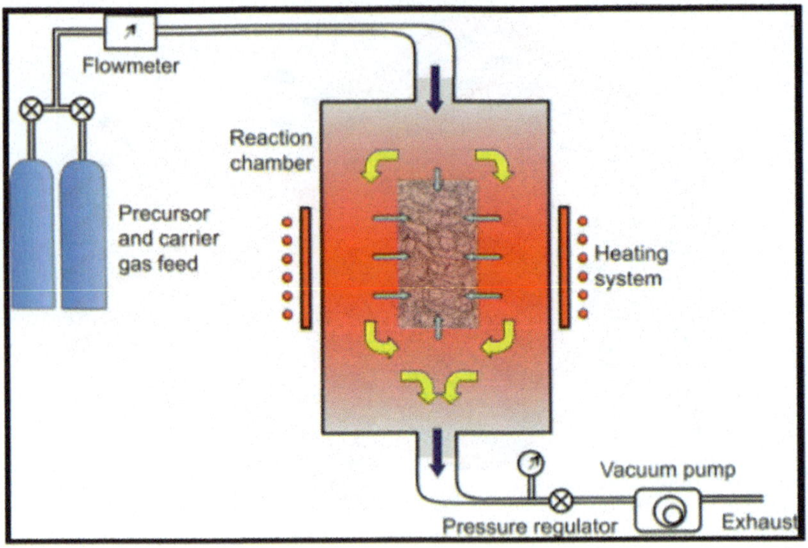

Fig. 4.7 Schematic of CVI process

transfer process, e.g., diffusion or convection, and adheres to the surface of the porous preform and the pore wall, and the reactant adsorbed on the wall undergoes a surface chemical reaction to generate solid products and release gas by-products, which are de-adhered from the wall surface, and enter into the main gas flow through the mass transfer process to be discharged out of the deposition furnace. Figure 4.7 shows the schematic of the CVI process. Figure 4.8 shows the CVI furnace used to fabricated the SiC/SiC combustor liner.

In the high temperature corrosive environment, the surface of the SiC/SiC composite will be oxidized to generate SiO_2. Due to the difference of the thermal expansion coefficient between the SiC fiber and the SiC matrix, the cooling holes of the SiC/SiC combustor liner are easy to crack during operation and thus seriously affect the durability and life of the combustor liner. The surface of the SiC/SiC combustor liner needs to be coated with EBC (Environmental Barrier Coatings, EBC), to improve the reliability and durability.

4.5 Machining of CMC Combustor Liner

CMC is a typical difficult-to-machine material, in addition to the characteristics of anisotropy, its hardness is second only to diamond and cubic boron nitride. Pulsed laser micromachining technology is used to realize the precise processing of SiC/SiC combustor liner micro-holes, and ultrasound-assisted machining technology

Fig. 4.8 CVI furnace for the fabricate the SiC/SiC combustor liner

is used to meet the demand for high-precision profile processing for various air-film holes with diameters larger than 0.5 mm, and the roughness can reach Ra = 0.3~0.5. Figure 4.9 shows the SiC/SiC combustor inner and outer liners after precise processing.

Fig. 4.9 SiC/SiC combustor inner and outer liners after precise processing

4.6 Assembly of CMC Combustor Liner

After the inner and outer liners and other components have been densified and processed, the SiC rivets are used for on-line riveting, which achieves a connection result equivalent to "metal riveting" after CVI-SiC deposition, improves the ability of the components to maintain the profile, and prevents delamination of the combustor liner during the service life. Figure 4.10 shows the SiC/SiC combustor liner after riveting. Figure 4.11 shows the CT diagram of the SiC/SiC combustor liner.

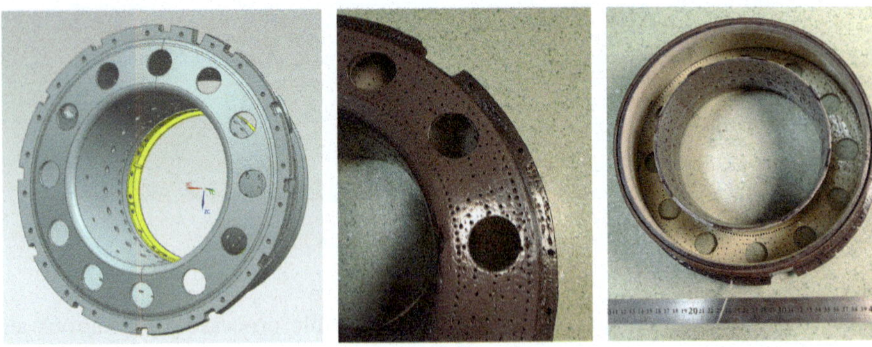

Fig. 4.10 SiC/SiC combustor liner after riveting

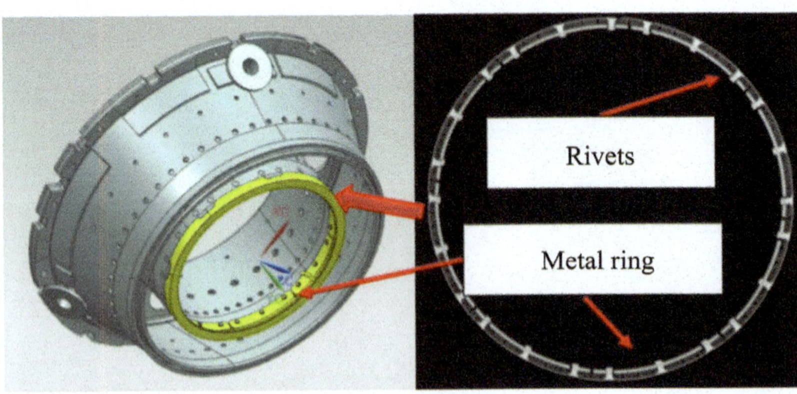

Fig. 4.11 CT diagram of the SiC/SiC combustor liner

4.7 Ground Testing of CMC Combustor Liner

On October 14, 2022, the ground ignition and performance test of the combustor liner was conducted, and the test state parameters are shown in Tables 4.3 and 4.4. Figure 4.12 shows the SiC/SiC combustor liner before and after ground tests. The average gas temperature was 1500 °C, with localized hot spots reaching 1800 °C. Localized EBC coating peeling was found after the ground testing.

Table 4.3 The first ground performance test parameters

Number	Inlet pressure P_{t3}/(MPa)	Inlet temperature, T_{t3}/(K)
1	0.378	495.27
2	0.567	562.34
3	1.137	645.2
4	1.389	680.71

Table 4.4 The second ground performance test parameters

Number	Inlet pressure P_{t3}/(MPa)	Inlet temperature, T_{t3}/(K)
1	1.137	645.20
2	1.520	700.00
3	1.610	713.95

Fig. 4.12 SiC/SiC combustor liner after ground test

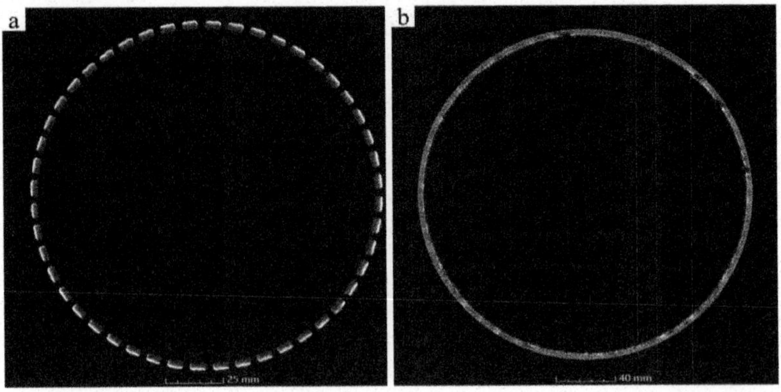

Fig. 4.13 Localized CT cross section of SiC/SiC combustor liner **a** inner liner; and **b** outer liner

4.7.1 EBC Wall Thickness Analysis of SiC/SiC Combustor Liner

Figure 4.13 shows the localized CT cross-section of SiC/SiC combustor liner. The EBC coatings covering the outer wall of the inner liner and the inner wall of the outer liner are denser compared to the CMCs, and the CT images of the EBC coatings show higher brightness.

Figure 4.14 shows the wall thickness distribution of the inner liner. The two-dimensional and three-dimensional views are shown at three angles. On the left side is the EBC color scale, where the transition from purple to red represents the gradual increase of the wall thickness from 0 to 1 mm. The EBC as a whole shows green color with high thickness uniformity, and it can be found that the EBC is a little thicker near the holes of the air film. After the ground test, the EBC of the inner liner exhibited localized peeling off. The EBC wall thickness is mainly concentrated around 0.47 mm, with more distribution around 0.51 and 0.60 mm.

Figure 4.15 shows the wall thickness distribution of the outer liner. The EBC of the outer liner was also peeling off in a few large areas, and the overall color of the coating exhibited a uniform cyan color that was thinner than the inner liner coating, with a slightly thicker coating near the air film holes. The wall thickness of the outer liner coating is mainly distributed in the range of 0.36 to 0.47 mm, which is more inclined to smaller values compared to the inner liner coating.

4.7.2 Porosity Analysis of SiC/SiC Combustor Liner

Under quantitative characterization of the pores in the outer liner, the resolution of the CT scan was 0.09 mm, and the equivalent diameter of the inner liner pores was a minimum of 0.53 mm, with a maximum value of 14.60% achieved at 0.58 mm and

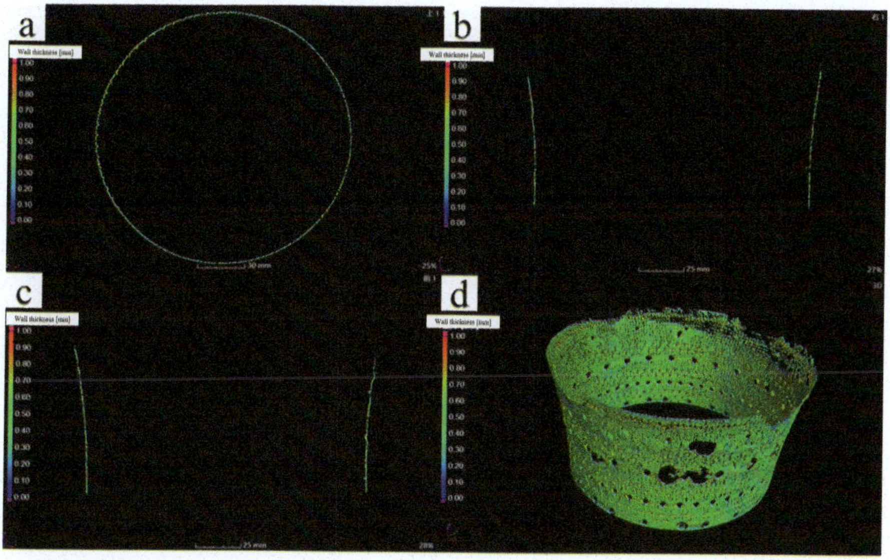

Fig. 4.14 The wall thickness distribution of inner liner

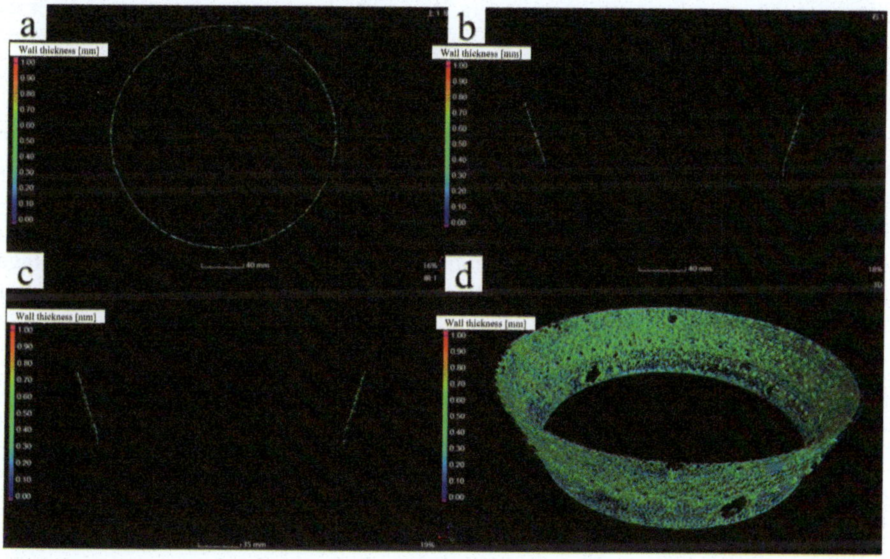

Fig. 4.15 The wall thickness distribution of outer liner

Fig. 4.16 Large pores with equivalent diameter greater than 3 mm in the inner liner

0.65% at 2.13 mm pore diameter. The pore surface area achieved a maximum value of 15.75% at 1.75 mm^2 and the volume achieved a maximum value of 19.39% at 0.13 mm^3. As the pore diameter increases, the pore occupancy ratio shows a tendency to increase and then decrease. The porosity of the SiC/SiC inner liner after the test is about 8.49%. The pores with equivalent diameters larger than 3 mm (defined as large pores) in the inner liner are given in Fig. 4.16. The large pores in the inner liner of the combustor liner are mainly concentrated around the air film holes, which are of different shapes and appear as irregular sheets. The largest number of large pores is found near the largest-size air film holes.

Quantitative characterization of the pores in the outer liner was conducted with a CT scan resolution of 0.07 mm, and the minimum value of the equivalent diameter of the pores in the outer liner was 0.43 mm and achieved a maximum value of 24.42% at 0.48 mm, a maximum value of 19.72% of the surface area at 0.90 mm^2, and a maximum value of 19.39% of the volume at 0.05 mm^3. The pore number percentage increases and then decreases as the scale increases. The porosity of the SiC/SiC outer liner after the test is about 11.35%. The pores with equivalent diameters greater than 1.5 mm were visualized in three dimensions, and the main form of defects in the outer liner after the test was delamination, which is indicated in red in Fig. 4.17. The locations of the delamination defects were concentrated at the bottom of the outer liner, located below the corresponding label sheets: 5–6, between 7–8 and 11–12, and between 21–22 and 23–24, due to the large temperature difference between the inner and outer surfaces of the outer liner during the test.

4.8 Summary and Conclusions

In this chapter, the design, fabrication and testing of the CMC combustor liner were investigated. The SiC/SiC combustor liner was fabricated using the CVI process, and coated with Environmental Barrier Coatings (EBC). Pulsed laser micromachining

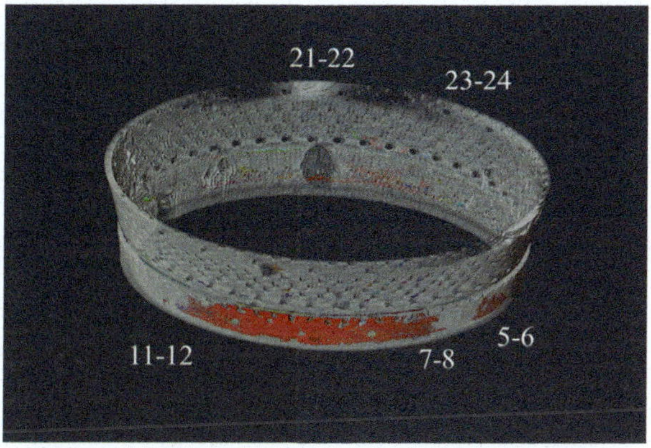

Fig. 4.17 Large pores with equivalent diameter greater than 1.5 mm in the outer liner

technology was used to realize the precise processing of SiC/SiC combustor liner micro-holes, and ultrasound-assisted machining technology was used to meet the demand for high-precision profile processing for various air-film holes with diameters larger than 0.5 mm. After the inner and outer liners and other components were densified and processed, the SiC rivets were used for on-line riveting. On October 14, 2022, the ground ignition and performance test of the combustor liner was conducted. The average gas temperature was 1426 °C, with localized hot spots reaching 1734 °C. Localized EBC coating peeling was found after the ground testing. After the ground testing, the EBC, pores of the inner and outer liners were analyzed using the CT scan.

(1) The overall distribution of SiC/SiC combustor liner EBC was uniform, the outer liner coatings were slightly thinner than the inner ring coatings, and the EBC thicknesses were distributed between 0.3 and 0.6 mm.
(2) The porosity of the inner liner was 8.49%, and the large-size pores were mainly concentrated near the largest-size air-film holes, which may be triggered by the interconnection and expansion of small pores triggered by machining.
(3) The porosity of the outer liner was 11.35%, and the EBC of the outer liner was locally peeled off, and the delamination phenomenon of the outer liner occurred locally after the test, which is concentrated in the thicker area at the bottom of the outer liner.

References

Bhatia T, Jarmon D, Shi J et al (2010) CMC combustor liner demonstration in a small helicopter engine. In: ASME turbo expo 2010: power for land, sea, and air. American Society of Mechanical Engineers Digital Collection, Glasgow, pp 509–513.

Lacombe A, Spriet P, Alain A, Bouillon E, Habarou G 2009 Ceramic matrix composites to make breakthroughs in aircraft engine performance. In: 50th AIAA, ASME, ASCE, AHS structural dynamic, and materials conference. Palm Springs, California, USA, pp 4–7

Li L (2021a) Strain response of fiber-reinforced ceramic-matrix composites subjected to stress-rupture with stochastic load at intermediate temperature. J Compos Mater 55(2):3035–3048

Li L (2021b) Micromechanical modeling of cyclic non-closure hysteresis loops of fiber-reinforced ceramic-matrix composites considering variable matrix fragmentation density. Mater Sci Eng, A 805:140795

Li L (2021c) High temperature mechanical behavior of ceramic-matrix composites. Wiley-VCH, Weinheim, Germany

Li L (2022) Hysteresis-based identification approach for crack opening and closure stress in SiC/SiC fiber-reinforced ceramic-matrix composites. Int J Fatigue 162:106945

Li L (2023a) Hysteresis-based approach to estimate constituent properties of SiC/SiC ceramic matrix composites. Int J Fatigue 167:107302

Li L (2023b) Ceramic matrix composites: lifetime and strength prediction under static and stochastic loading. Elsevier, Oxford, UK

Misra AK (2001) Development of advanced engine materials in NASA ultra efficient engine technology program. In: 15th international symposium on air breathing engines conference, Louisiana

Padture NP (2016) Advanced structural ceramics in aerospace propulsion. Nat Mater 15s:804–809

Chapter 5
Design, Fabrication and Testing of CMC Turbine Blisk

Abstract In this chapter, the design, fabrication and testing of CMC turbine blisk were investigated. The service performance analysis of the turbine blisk was at first analyzed. The spider web structure (SWS) fiber preform was adopted as the reinforcement elements of the turbine blisk. The BN interphase and SiC matrix were deposited on the SWS fiber preform, and during the fabrication, the on-line machining were conducted to control the shape and dimension of the blisk. The environmental barrier coatings (EBCs) were prepared on the surface of the SiC/SiC turbine blisk using the atmospheric plasma spraying (APS) method. The XCT scan was conducted to characterize the internal defects in the SiC/SiC turbine blisk. The mechanical performance evaluation, over rotation test and engine bench test were performed, respectively. Finally, the SiC/SiC turbine blisk was installed on an aircraft and completed the first flight testing, which verified the feasibility of the SiC/SiC turbine rotor for gas turbine engine application and accumulated valuable experience for the subsequent development of CMC hot-section components.

Keywords SiC/SiC · Turbine blisk · Spider web structure (SWS) · XCT · FEM · Ground testing · Flight testing

5.1 Introduction

To improve launch vehicle safety and reliability, and reduce space transportation costs, NASA has worked to develop CMCs for turbine blisk of rocket engine (Bhat 2018). Figure 5.1 shows the turbine blisk for rocket engines developed at NASA Marshall Space Flight Center. The Center has conducted dozens of tests on the monolithic turbine blisk, and accumulated a large amount of experimental data to predict the lifetime of CMCs in rocket engines. The material of monolithic turbine blisk was 2D C/SiC with a disk diameter of 7.6 inches (19 cm) and a thickness of approximately 3/4 inch (2.0 cm) and was manufactured by Honeywell Advanced Composites. The turbine blisk was tested at 25,000 r/min. Studies have shown that this CMCs turbine blisk can withstand temperatures of 2,000°F (1,093 °C), much

X. Liu and L. Li, *Design, Fabrication and Testing of Aeroengine Ceramic-Matrix Composite Components*, Advanced Ceramics and Composites 7,
https://doi.org/10.1007/978-981-97-6109-8_5

Fig. 5.1 2D C/SiC turbine blisk developed by NASA marshall space fight center

higher than the 1,200°F (649 °C) that the nickel alloy turbine blisk can withstand. Clearly when a rotating turbine powers an engine, the more heat it can withstand, the less fuel is required; the increased engine efficiency allows heavier payloads to be boosted to higher orbits. In addition, the payload can be increased because the CMCs turbine blisk is only about a quarter the weight of a metal blisk, and the CMCs monolithic turbine blisk is particularly resistant to damage, in addition to being lightweight and heat resistant. The tests found that the turbine blisk continued to work properly despite a crack occurred in one of the blades. Engineers performed millions of cycles on the turbine blisk, proving the outstanding durability under stress.

NASA's Lewis research center explored the feasibility of CMCs turbine rotors for liquid rocket engine applications. Figure 5.2 shows the C/SiC turbine blisk developed by NASA Lewis research center. The study showed that C/SiC turbine blisk can maintain high strength at 1370 °C, which is 440 °C higher than the heat-resistant temperature of high-temperature alloys, and have improved vibration damping, which is 10 times higher than that of metal components, and can effectively mitigate mechanical vibration fatigue problems. This study shows that CMCs have vibration absorption and damping properties, which are expected to improve the structural stability and service safety of turbine rotors.

The Japan Aerospace Exploration Agency (JAXA) has developed a three-dimensional woven C/C turbine blisk with inner and outer diameters of 180 and 200 mm, respectively (Ogawa et al. 2001). The rotation speed reached 6000 r/min (blade tip speed of 600 m/s) at room temperature. The simulation analysis found that the stress at the center of the turbine blisk was the highest. To increase the turbine blisk strength, the volume content of the radially fibers was increased at turbine blisk $r > 130$ mm, and in fact the turbine blisk rupture occurred at the fiber discontinuity region. Figure 5.3a shows the moment when the 3D-C/C turbine blisk was damaged, and the fiber breakage occurred in the middle section of the turbine blisk, due to excessive internal stress at the location of the radially increased fibers. Figure 5.3b shows a photograph of the 3D-C/C turbine disk, which has eight circular protrusions on the outer edge of the turbine blisk as the load transfer interface between the

Fig. 5.2 C/SiC turbine blisk developed by NASA's Lewis research center

rotating shaft and the turbine blisk. This interface can reduce the load transfer stress between the turbine disk and the rotating shaft, and the thickening protrusions can significantly reduce the circumferential tensile stress around the hole.

The Rolls-Royce company provides a method for preparing a CMC turbine blisk, where the turbine blisk includes a plurality of individual turbine blades, a turbine disk outer ring and a turbine disk (Freeman and Engel 2019). The turbine blades are fixed to the outer edge of the outer ring of the turbine disk, and the structure is schematically shown in Fig. 5.4.

For small gas turbine engines, the Russian Central Institute of Aviation Motors (CIAM) has developed a series of CMCs turbine stationary and rotor components, which are currently being tested and verified on small gas turbine engine platforms (Nozhnitsky 2018). Figure 5.5 shows the turbine rotor components developed by CIAM, using the development concept of separate system preparation and combined assembly. The advantages of this solution are mainly: high blade strength and structural adaptability, which can be applied to different sizes of aeroengines.

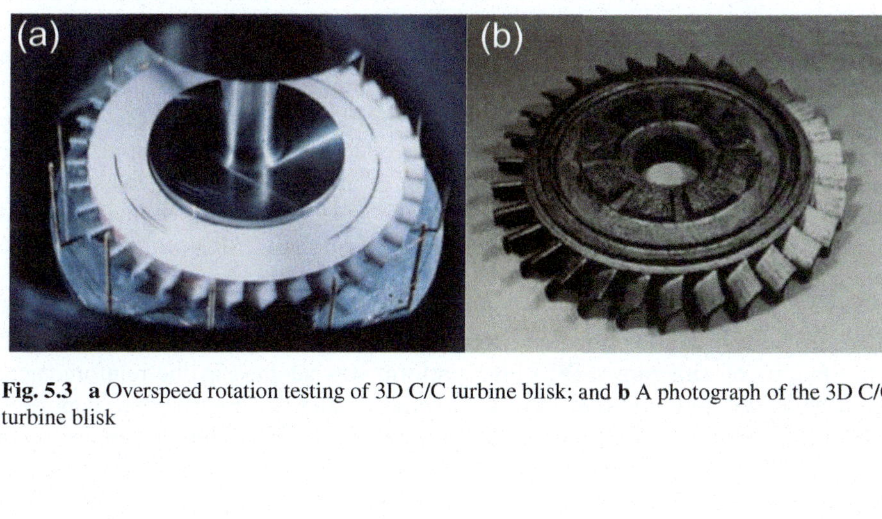

Fig. 5.3 **a** Overspeed rotation testing of 3D C/C turbine blisk; and **b** A photograph of the 3D C/C turbine blisk

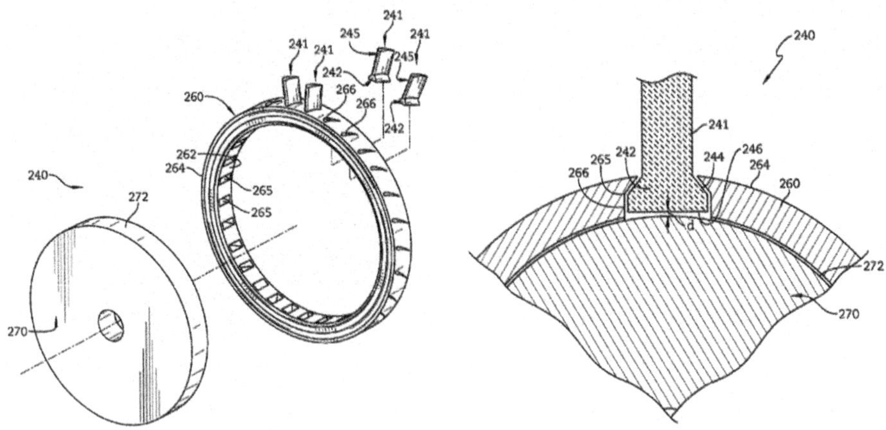

Fig. 5.4 Schematic of CMC turbine blisk developed by Rolls-Royce company

Fig. 5.5 CMC turbine rotor components developed by CIAM

On February 10, 2015, GE company introduced SiC/SiC rotating parts to the harshest aeroengine operating conditions for the first time, demonstrating and verifying SiC/SiC low-pressure turbine blades on the F-414 turbofan engine with complete success, which represents a major technological breakthrough in the jet propulsion field. The SiC/SiC turbine blades tested by GE, shown in Fig. 5.6, were also developed using a split system and integrated assembly; the central disc was made of metal and the high-temperature part of the outer edge was made of SiC/SiC turbine blades; some of the blades were coated with a yellow environmental barrier coating (EBCs). GE believes that the application of CMCs will revolutionize jet engines.

In this chapter, the design, fabrication and testing of CMC turbine blisk were investigated. The service performance analysis of the turbine blisk was at first analyzed. The spider web structure (SWS) fiber preform was adopted as the reinforcement elements of the turbine blisk. The BN interphase and SiC matrix were deposited on the SWS fiber preform, and during the fabrication, the on-line machining were

Fig. 5.6 SiC/SiC turbine blade developed by GE company

conducted to control the shape and dimension of the blisk. The environmental barrier coatings (EBCs) were prepared on the surface of SiC/SiC turbine blisk using the atmospheric plasma spraying (APS) method. The XCT scan was conducted to characterize the internal defects in SiC/SiC turbine blisk. The mechanical performance evaluation, over rotation test and engine bench test were performed, respectively. Finally, the SiC/SiC turbine blisk was installed on an aircraft and completed the first flight testing.

5.2 Service Performance Analysis of Turbine Blisk

Turbine blisk in service environment possess high rotation speed, the root of the inner hole in the blisk and the root of the blade produce large circumferential and radial stresses, respectively, leading to the high performance requirements of CMCs. An aeroengine turbine blisk has a small diameter (Φ84.5), high speed (100,000 rpm and above), high service environment temperature, along with vibration and other extreme environmental conditions, and the turbine blisk service life is long. The study shows that the stress distribution of the blisk root and blade root under 112,000 rpm rotating condition is more concentrated, the stress is at the range of 200–300 MPa. Figure 5.7 shows the FE simulation results.

5.3 Design of CMC Turbine Blisk

Figure 5.8 shows the loading condition of the turbine blisk under service conditions. The turbine blisk will be subjected to two main stresses, i.e., tangential stress and centrifugal stress. Compared with the homogeneous material, the structure and function of composite material can be designed to be stronger; this study proposes to design and arrange the reinforcing fibers in the annular and radial directions of the

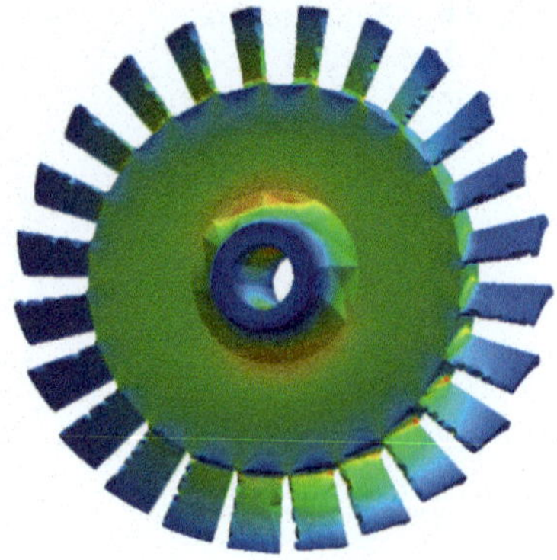

Fig. 5.7 Stress analysis of CMC turbine blisk at 112,000 rpm

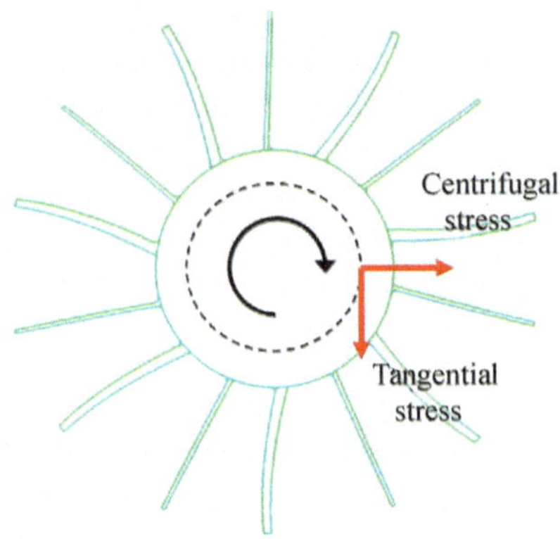

Fig. 5.8 Diagram of stress distribution of turbine blisk under working condition

turbine blisk to carry the load, and build a turbine rotor structure of CMCs suitable for the turbine rotor working conditions.

It has been found that spider webs in nature consist of annular or spiral annular spider silk and ray-like radial spider silk; spider webs can effectively resist different loads, such as wind loads and impact loads. Even, the spider web can guarantee the spider to successfully complete the prey hunting under the condition of partial damage of the spider web. Inspired by the spider web, it is a logical starting point for this study to see if the high strength and reliability of the spider web can be "transferred" to the turbine rotor components. Cranford et al. (2012) suggest that this characteristic of spider webs is not only due to the toughness of the spider silk, but also closely related to the geometric configuration of the spider web; the unique "concentric circle + ray-like radial spider silk" configuration of the spider web can tolerate the local fracture of the spider structure at the stress concentration region, thus avoiding further propagation of the damage. In addition, when the spider web is locally damaged, the external load will be redistributed, which can avoid the system reconstruction due to the damage of the whole system. From the bionic point of view, the structural features of the spider web will help to improve the reliability and load-bearing capacity of the turbine rotor. Based on the spider web configuration, we try to "transfer" the characteristics of spider web such as impact resistance, local damage resistance and overall synergistic load-bearing to CMCs turbine blisk components, so that the CMCs turbine blisk can avoid the defects of conventional materials, and improve the service performance of turbine blisk.

The spider web structure (SWS) preform unit layer structure design is shown in Fig. 5.9. The weft fibers are arranged in concentric circles or spirals with the center of the disk as the origin, and the radial fibers are distributed within the radial plane. As the distance between the fibers and the origin gradually increases, the area within the adjacent radial fiber angle will also gradually increase, and then it is necessary to add yarn at the appropriate position inside the adjacent radial fiber angle, aiming to increase the local fiber volume content to ensure the uniform fiber content within the unit layer and reduce the internal stress concentration of the material. Moreover, the weft and warp fibers are interwoven and shaped at their intersection points. Figure 5.10 shows the cross-section view of the SWS preform unit. The fiber preform adopts the SiC fibers with fiber bundle size 0.5 k, the fiber monofilament diameter is 14–16 μm, and density $\rho = 2.75$ g/cm^3. Figure 5.11 shows the schematic diagram of the SWS fiber preform. The SWS fiber preform is formed by gradually laminating the unit layers to reach the designed thickness and then pressing and shaping it with a mold. After molding the preforms, the laminated unit layers are then stitched together by a Z-directional stitching process to increase the interlayer bonding strength of the final composite. The stitching spacing is adjustable from 10 to 30 mm depending on the thickness of the turbine blisk.

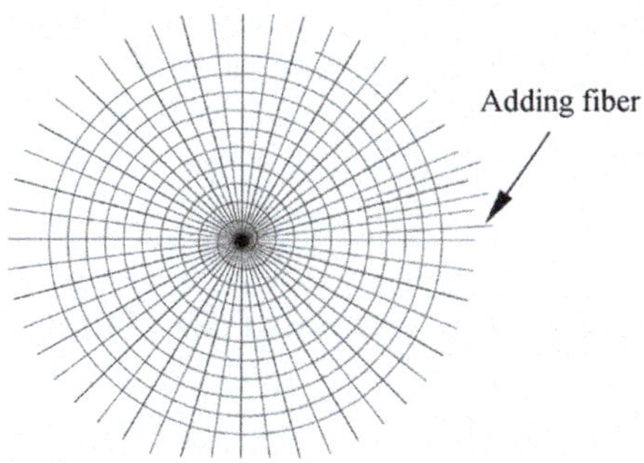

Fig. 5.9 Schematic diagram of spider web structure preform unit

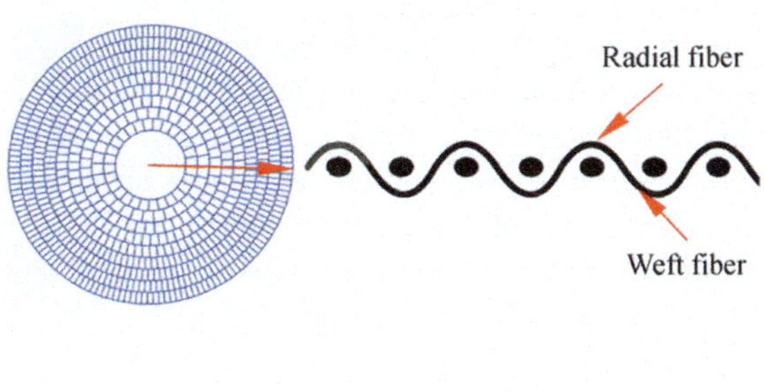

Fig. 5.10 Cross-section of SWS preform unit

Fig. 5.11 SWS fiber perform of SiC/SiC turbine blisk

5.4 Fabrication of CMC Turbine Blisk

5.4.1 Deposition of BN Interphase

The interface is the key to regulate the interrelationship between fibers and matrix and thus improves the mechanical properties of the composites, and its composition and structure play a crucial role in the effectiveness of the interface (Li 2020). The SiC/SiC interface layer is generally a pyrolytic carbon (PyC) or boron nitride (BN) interphase. In the present analysis, the long-life BN interphase was deposited on the SiC fiber surface. The monolithic turbine blisk preform was placed in the CVI deposition furnace with the vacuum level maintained at 10^{-3} Pa, and the temperature was increased to 1000–1200 °C; the preform BN interface layer was prepared with NH_3 as the nitrogen source, BCl_3 as the boron source, and H_2 as the dilution gas. The volume flow rates of NH_3, BCl_3, and H_2 were 200, 300, and 800 ml/min, respectively. The BN interphase thickness was about 500 nm, as shown in Fig. 5.12.

5.4.2 Deposition of SiC Matrix

The SiC matrix was prepared by CVI process on the surface of SiC fibers with the deposited BN interphase. Firstly, the SiC fiber preform was placed in the CVI furnace, and the air pressure inside the furnace was kept at $p = 20\sim50$ kPa, CH_2Cl_2Si was the SiC precursor, H_2 and Ar were the diluting gases (H_2:Ar volume ratio of 20:1), and the gas mixture flow rate was $f = 2000\sim3500$ ml/min, and the CH_2Cl_2Si gas was introduced into the reaction zone inside the CVI furnace. The furnace temperature was $T = 1000\sim1100$ °C, and single furnace deposition time was $t = 80\sim150$ h. Multiple cycles of densification deposition of SiC matrix were conducted. When the composite density $\rho \geq 2.5$ g/cm^3, the densification of the matrix was completed. The schematic diagram of SiC matrix densification is given in Fig. 5.13. Figure 5.13a

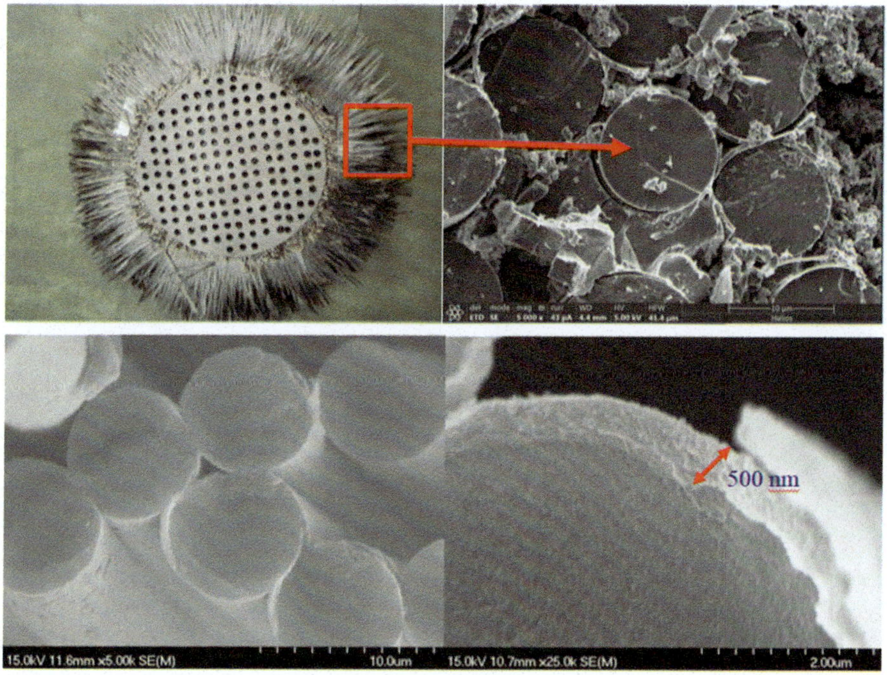

Fig. 5.12 Deposition of the BN interphase on the fiber preform

gives the state of the fiber preform after deposition of the SiC matrix, and Fig. 5.13b shows the state of SiC preform after high temperature mold removal, and the preform in this state already has rough processing conditions. After the matrix deposition, the SiC deposition purity is checked, as shown in Fig. 5.14. High quality SiC matrix deposition can improve the mechanical properties of SiC/SiC composites and enhance the structural strength of turbine blisk. As can be seen in Fig. 5.14, the deposited matrix is mainly SiC with a small amount of O and B.

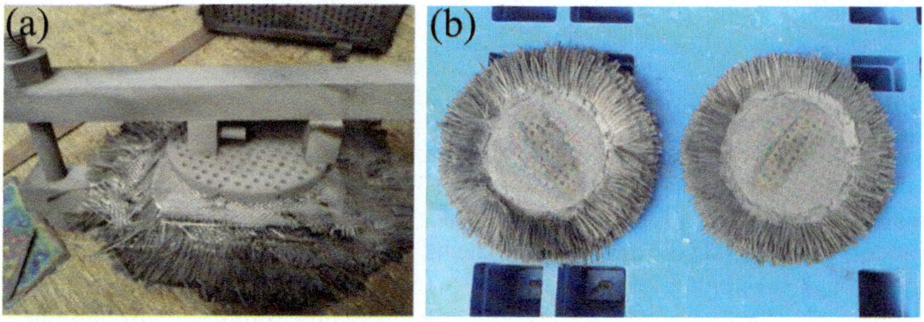

Fig. 5.13 SiC matrix densification on the SiC fiber preform **a** Deposition of SiC matrix; and **b** SiC preform after removing mold

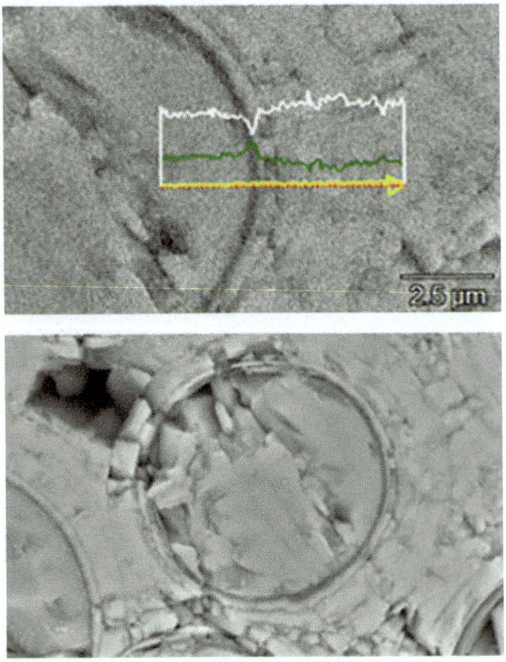

Fig. 5.14 Purity test result of the SiC matrix

5.5 Molding and Protection of CMC Turbine Blisk

The hardness of SiC matrix is second only to diamond and cubic boron nitride, and the SiC/SiC turbine blisk is a typical super hard material processing. The "on-line processing" method is used, and the mechanical processing and matrix deposition are carried out simultaneously and promote each other's process (Liu et al. 2023, 2022). The initial densification of the preform is machined, and the "hard shell" formed on the surface of the component by the matrix deposition can be opened during the near-size molding process, which is conducive to the subsequent further matrix deposition.

5.5.1 Rough Machining

After the turbine blisk density $\rho \geq 2.0\ \text{g/cm}^3$, the whole turbine blisk was machined on a multi-axis Computer Numerical Control (i.e., CNC) machining, and the machining process tried to avoid the destruction of the fiber preform structure. The turbine blisk after rough machining is shown in Fig. 5.15. This stage of rough machining not only enables the component to be formed in close dimensions, but also eliminates the "bottleneck" effect of the CVI process and promotes the densification process

Fig. 5.15 Turbine blisk after rough machining

of the matrix. Subsequently, the rough machined blisk are placed back into the CVI deposition furnace to continue the matrix densification.

5.5.2 Secondary Rough Machining

After rough machining, the blisk continues to be deposited for a certain number. When the density $\rho \geq 1.8$ g/cm^3, the secondary rough machining starts according to the preparation process. After this machining, the turbine disk size is basically with a margin of 0.5 mm reserved for the main size. Figure 5.16 shows the turbine blisk after the secondary rough machining.

5.5.3 Precision Machining

After the turbine blisk density $\rho \geq 2.0$ g/cm^3, the turbine blisk was machined on a multi-axis CNC machine tool. After the turbine blisk density $\rho \geq 2.4$ g/cm^3, the surface of the turbine blisk was deposited 50–200 μm SiC coating to seal the surface defects. Figure 5.17 shows the turbine blisk after precision machining.

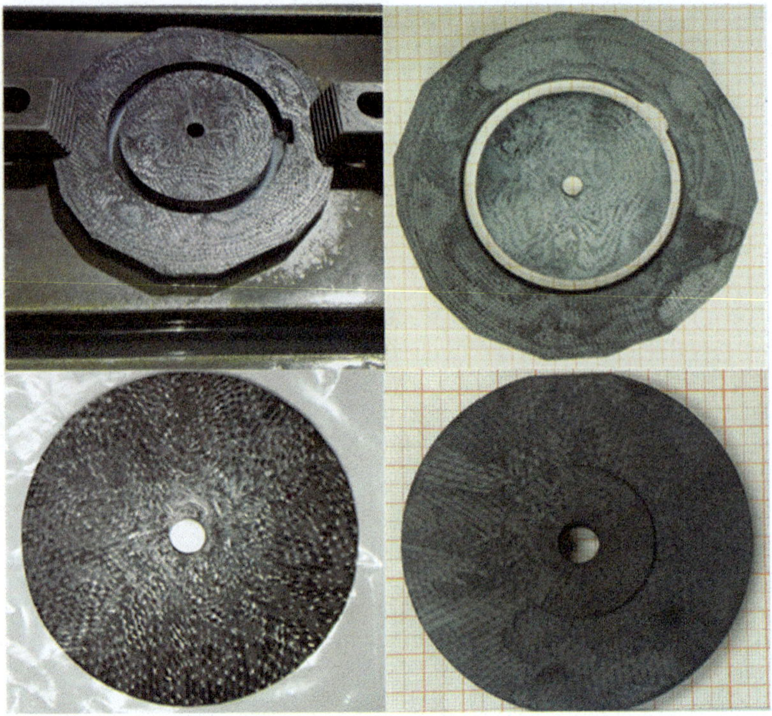

Fig. 5.16 Turbine blisk after secondary rough machining

Fig. 5.17 Turbine blisk after precison machining

5.5.4 Preparation of EBCs

The SiC matrix has good oxidation resistance, but defects and cracks inside the matrix will lead to oxygen entering inside the composite and oxidizing the interphase and fibers. The environmental barrier coating (EBCs) was prepared by the Atmospheric Plasma Spraying (APS) method.

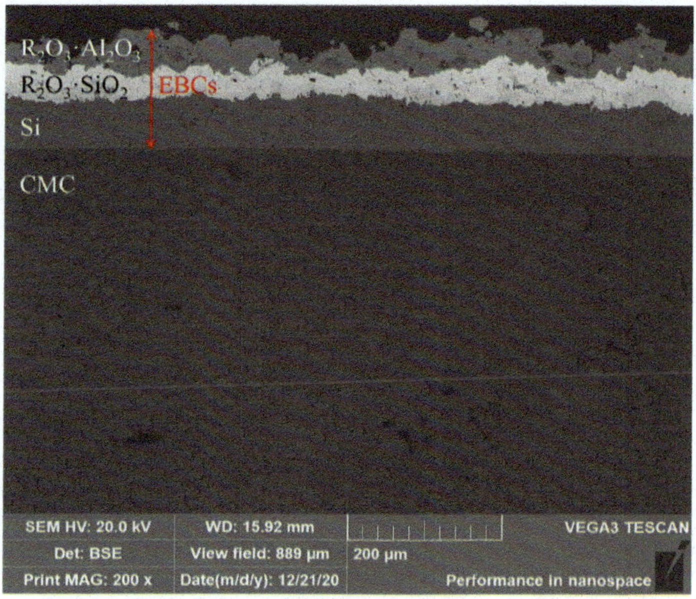

Fig. 5.18 Cross-section SEM morphology of EBCs

Figure 5.18 shows the structural morphology of EBCs, which contains three functional layers, as following:

(1) Antioxidant Si layer on the surface of CMC;
(2) Intermediate layer $R_2O_3{\cdot}SiO_2$ to resist hydroxide corrosion;
(3) Thermal insulation and antihydroxide corrosion surface layer $R_2O_3{\cdot}Al_2O_3$, where R = La/Nd/Sm/Gd/Yb.

The thickness of each layer is 50~100 μm, and the total thickness does not exceed 300 μm.

5.6 X-ray CT Scanning of CMC Turbine Blisk

In addition to the conventional dimensional accuracy and appearance quality, the size and distribution of the internal defects of the turbine blisk have a significant impact on the overall performance of the turbine blisk. Figure 5.19 shows the X-ray nondestructive testing of the turbine blisk after rough machining, secondary rough machining and precision machining. It is found that the pores inside the turbine blisk are still large after rough machining and the uniformity of the material is poor; while the denseness of the finished turbine blisk is high and the large pores inside the material have disappeared after several depositions, and the overall uniformity of the disk material has been significantly improved.

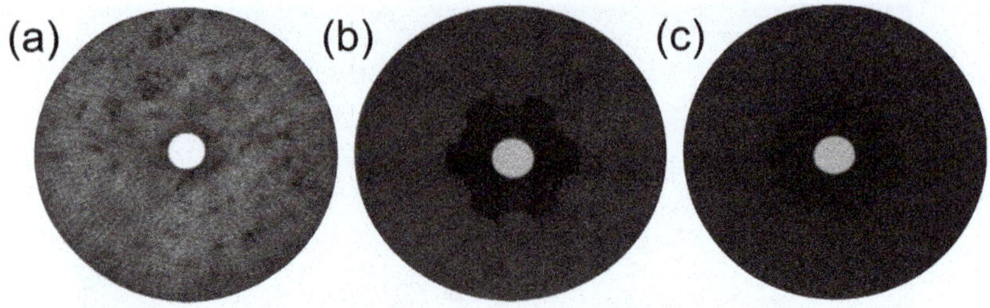

Fig. 5.19 X-ray CT scan of turbine blisk **a** after rough machining; **b** after second rough machining; and **c** after precision machining

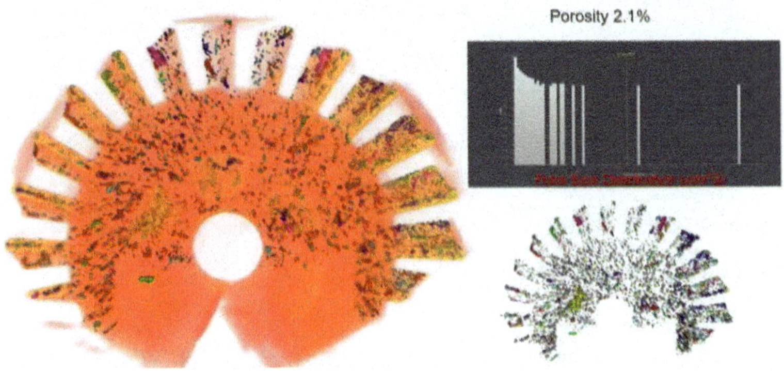

Fig. 5.20 3D reconstruction analysis of SiC/SiC turbine disk CT data

To characterize the distribution of micro-defects inside the fabricated turbine blisk, the 3D reconstruction imaging was used to realize the whole and local for defect detection of SiC/SiC turbine blisk, as shown in Fig. 5.20. It was found that:

(1) The largest number of porosity defects smaller than 30 μm accounted for 11% of the porosity defect data.
(2) The porosity defects of 40 μm accounted for 5% of the porosity defect data.
(3) The porosity defects larger than 100 μm accounted for about 3%.

From the distribution of various types of defects, defects smaller than 40 μm are mainly distributed in the disk area, while 60% of the large defects larger than 40 μm are distributed in the blade or blade tip position.

5.7 Assembly of CMC Turbine Blisk

Figure 5.21 shows the machining of the metal clutch sleeve for the turbine disk and the overall situation after the combined assembly with the turbine disk. The metal clutch sleeve plays the function of load transfer and stress reduction of SiC/SiC

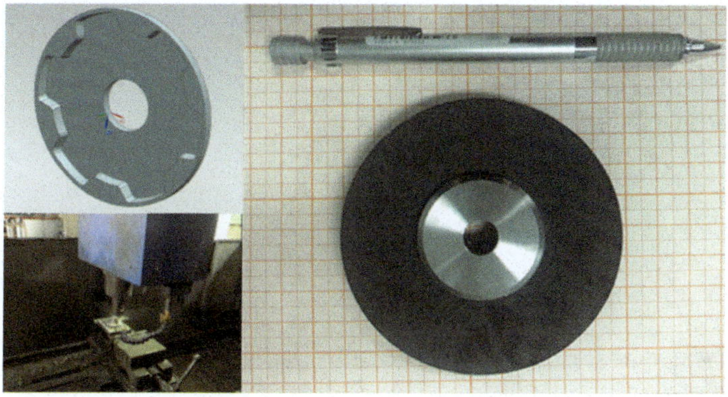

Fig. 5.21 Machining of metal clutch sleeve and assembly of SiC/SiC turbine blisk

turbine blisk under actual working condition. A single turbine disk with two metal clutch sleeves was tested in stainless steel at room temperature, and a metal clutch sleeve prepared from GH4169 alloy was used for the high temperature test.

5.8 Performance Evaluation of CMC Turbine Blisk

5.8.1 FE Stress Analysis

Figure 5.22a shows the results of FE stress analysis of SiC/SiC turbine blisk with a rotation speed of 100,000 rpm. The simulation results show that the high stress area of the turbine blisk is mainly concentrated at the position of the inner hole, with the maximum stress value of 164 MPa; the minimum stress area is at the tip of the turbine blisk, with about 3.4 MPa, indicating that the centrifugal stress and shear stress caused by the high-speed rotation of the turbine blisk will trigger the outward expansion of the inner hole of the disk, resulting in a large circumferential tensile stress. Therefore, more reinforcing fibers need to be arranged at the inner hole of the turbine disk for load bearing, or other processes need to be adopted to increase the fracture strain and matrix cracking stress of the composite material at the inner hole of the turbine disk.

Figure 5.22b shows the strain analysis of SiC/SiC turbine blisk at 100,000 rpm using the FE method. The simulation results show that the high strain area of the turbine blisk is mainly concentrated at the position of the inner hole, with the maximum deformation value of 0.0005 mm; the minimum stress area is at the tip of the turbine blisk, with the deformation of about 9.0×10^{-6} mm.

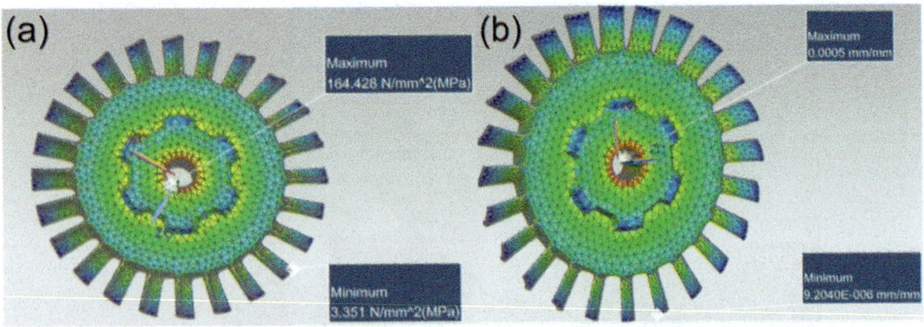

Fig. 5.22 FE analysis of SiC/SiC turbine blisk at 100,000 rpm **a** Stress distribution; and **b** Strain analysis

5.8.2 FE Strength Prediction

The strength of SiC/SiC turbine blisk and its verification method were established. Firstly, a circular specimen was selected on the outer side of the SiC/SiC turbine blisk using the method of sampling the turbine blisk, as shown in Fig. 5.23a; and the notches were machined on both sides of the horizontal side of the circular ring, as shown in Fig. 5.23b; and the ultimate fracture load of SiC/SiC ring specimens were obtained by loading tests, as shown in Fig. 5.23c and d. The ultimate fracture strength of the turbine blisk was obtained as $\sigma_{uts} = 301$ MPa using FE simulation, as shown in Fig. 5.23e. The rupture speed of the SiC/SiC turbine blisk was obtained by the FE simulation, and the stress distribution of the SiC/SiC turbine blisk at a speed of 148,000 rpm is given in Fig. 5.23f. The maximum stress at the core hole is $\sigma = 301.95$ MPa, and the corresponding rupture speed of the SiC/SiC turbine blisk is 148,000 rpm.

5.9 Ground Testing of CMC Turbine Blisk

5.9.1 Overspeed Rotation Experiments at Room Temperature

The overspeed rotation experiments of the turbine blisk at room temperature was conducted on a high-speed rotor rotation test bench developed by the High Speed Rotating Machinery Laboratory of Zhejiang University. The test bench consists of a rotation mechanical system, electrical instrument and computer control system. The drive motor power is 45 kW, the maximum test rotor weight is 50 kg, the maximum test rotor diameter is $\phi600$ mm, and the over rotation test is performed under vacuum less than 200 Pa. The process data is collected through the computer system, and the high-speed photography equipment records the turbine blisk state in real time.

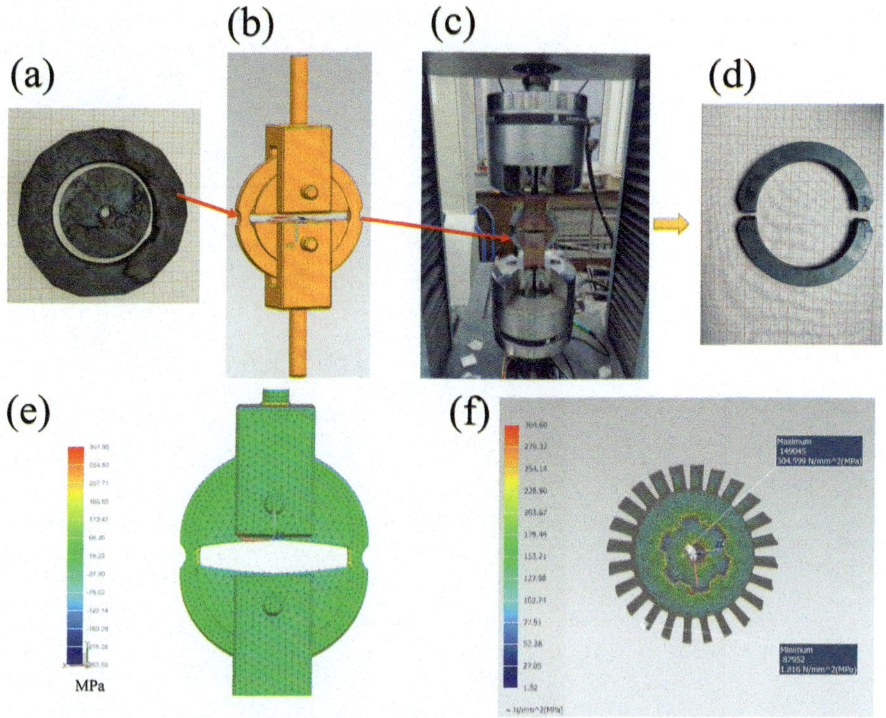

Fig. 5.23 Evaluation of SiC/SiC turbine blisk performance **a** Circular specimen; **b** Notched circular specimen; **c** Tensile fracture test; **d** Fracture specimen; **e** Finite element simulation of notched circular specimen fracture; and **f** Turbine blisk stress distribution

The expected rotational speed of the SiC/SiC turbine blisk was set to 60,000, 6,500, 70,000, 85,000, 100,000, 120,000 r/min, respectively, and the free mode test of the turbine blisk after each rotation test was conducted. Figure 5.24 shows a photograph of the turbine blisk cracking process recorded by the high-speed photographic equipment. The over-rotation test was conducted at room temperature with vacuum level of 150 Pa. The turbine blisk was damaged at 118,000 r/min. At T1 time, the turbine blisk was intact; at T2 time, the crack was found to exist from the edge of the inner hole of the blisk, and then the crack propagated along the meridian surface of the blisk to the outer edge of the blisk. Almost at the same time, the fracture of the blade occurred from the root of the blisk. At T6 time, the ruptured disk and blades were thrown out in all directions under the action of centrifugal force, and the turbine blisk was destroyed. The damage process of the turbine blisk is given as following,

(1) When the rotational speed reaches the limit, the edge of the inner hole of the turbine blisk cracked and gradually propagated along the meridian surface, which cut off the circumferential fibers
(2) At the same time, the stress at the blade root also reached the material limit, leading to the fracture of the blade root.

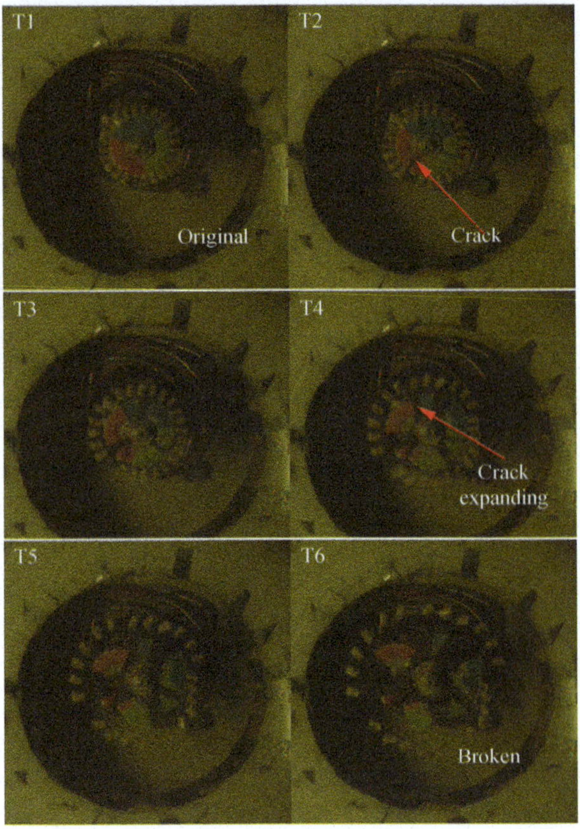

Fig. 5.24 Photo of high speed camera for test of SiC/SiC turbine blisk

(3) As the turbine blisk rotation speed increased, the stress growth in the inner hole and the blade root of the blisk is almost synchronized, leading to the simultaneously rupture of the blisk body and blade.

Figure 5.25 shows the morphology of the SiC/SiC turbine blisk before and after the damage. The damaged blisk has six penetrating cracks along the meridional surface, and the cracks all originate from the root of the bulge block at the hub location, indicating a high stress at this location. Most of the rotor blades also fractured from their roots. It indicates that the high stress area of the SiC/SiC turbine blisk is concentrated in the hub position and the outer blade root position.

Figure 5.26 shows the microstructure of the fractured blisk. A large number of fiber bundles and monofilament fibers were found to be pulled out successively in the disk section, and the pull-out of fracture fibers increased the energy dissipation during the rupture process.

Fig. 5.25 Pre and post burst condition of the SiC/SiC turbine blisk

Fig. 5.26 Fracture surface of fractured SiC/SiC turbine blisk

Modal testing is an effective means of non-destructive characterization of component damage, and the results of the SiC/SiC turbine blisk modal testing are shown in Table 5.1 for each test stage. The modal test was conducted using the "force hammer-acoustic pressure sensor" measurement method, and the data was collected in real time using an INV302 data acquisition analyzer. The first-order, second-order and third-order characteristic frequencies of the turbine blisk measured before the over-rotation test were $f = 10.97$, 15.85 and 21.65 kHz, respectively, which were much higher than the characteristic frequencies of the conventional metallic turbine blisk ($f \leq 5$ kHz). This was due to the higher stiffness of the spider web configuration and the fact that the specific stiffness of SiC/SiC was much higher than that of the metallic material. This helps to avoid coupling resonance between the turbine blisk and other components such as the drive shaft, which can improve the structural reliability of the aeroengine and increase the design margin of the components.

- After the over-rotation test at 60,000 r/min, the drops of the first, second, and third-order frequencies were 0.41, 0.03, and 0.26%, respectively. No obvious damage occurred in the blisk.

Table 5.1 Result of frequency test for SiC/SiC turbine blisk

Sample no	Speed/ ($r \cdot min^{-1}$)	Frequency/kHz			1-drop/%	2-drop/%	3-drop/%
		1*	2*	3*			
0	0	10.97	15.85	21.65			
1	60 000	10.92	15.84	21.59	0.41	0.03	0.26
2	65 000	10.96	15.84	21.68	0.05	0.03	−0.16
3	70 000	10.93	15.80	21.60	0.32	0.28	0.21
4	85 000	10.90	15.73	21.54	0.59	0.73	0.49
5	100 000	10.85	15.65	21.51	1.05	1.23	0.63

Note 1*—first order frequency, 2*—second order frequency, 3*—third order frequency

- After the over-rotation test at 85,000 r/min, the drops of the first, second, and third-order frequencies were 0.59, 0.73, and 0.49%, respectively.
- After the over-rotation test at 100,000 r/min, the drops of the first, second, and third-order frequencies were 1.05, 1.23, and 0.63%, respectively. Damages occurred in the blisk.

5.9.2 Aeroengine Bench Testing

The bench test can expose the problems in the design stage of the aeroengine and verify the performance and reliability of the SiC/SiC turbine blisk. The turbine blisk was assembled and the complete engine test run was carried out for verification, as shown in Fig. 5.27a and b.

The engine test condition was regulated and the engine fitted with the SiC/SiC turbine blisk was tested in accordance with the given speed spectrum. Figure 5.28 shows the speed, exhaust temperature and fuel consumption curves for the first cycle of the SiC/SiC turbine blisk. The SiC/SiC turbine blisk accumulated $N = 994$ cycles at maximum speed of $n = 60{,}000$ r/min and $N = 100$ cycles at maximum speed of $n = 70{,}000$ r/min.

The morphology of the original turbine blisk, the EBCs-SiC/SiC turbine blisk, and the turbine blisk after experimental tests are given in Fig. 5.27c-e. After undergoing $N = 994$ cycles at the highest speed of $n = 60{,}000$ r/min, the turbine blisk appearance shows a white bare material with powder state. The white bare material is $Si(OH)_4$ produced by SiC corroded by water oxygen, and the corroded material is in a glassy state, which is thrown out by high speed rotation and presents fluid phenomenon, and the blade part has high temperature and forms steam. The surface of the blisk tends to present as a river in the centrifugal direction, as shown in Fig. 5.27f, which is due to the pneumatic heating caused by high-speed rotation to oxidize the surface of the blisk, and the oxidation-generated material presents streaks of emission release under centrifugal force. Figure 5.27g and h give the SEM observation diagrams, the oxidized material is for the boron oxide formed under water-oxygen corrosion, and the boron element mainly originates from the BN interfacial phase on the SiC fiber surface, which indicates that the fiber surface interphase was oxidized during the test.

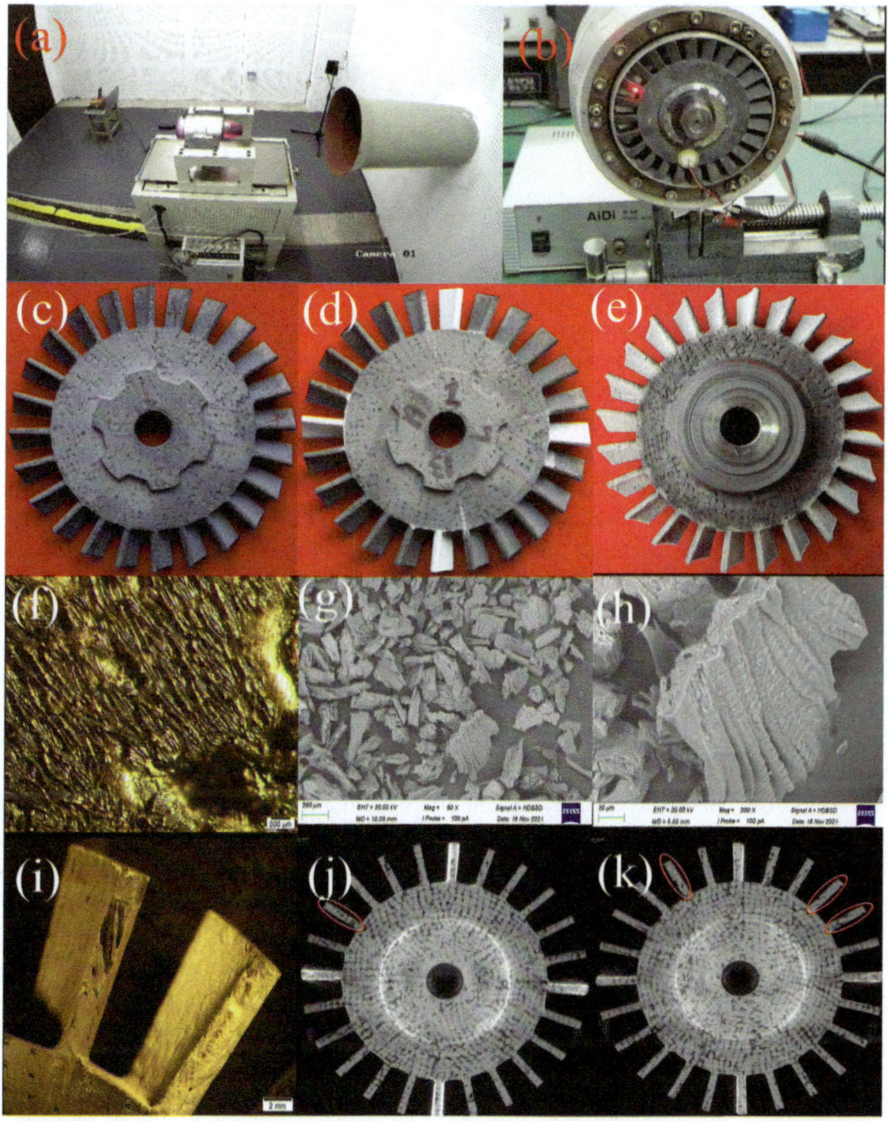

Fig. 5.27 Engine bench test assessment **a** Engine test bench; **b** SiC/SiC turbine blisk assembled in the test engine; **c** Original SiC/SiC turbine blisk; **d** EBCs-SiC/SiC turbine blisk; **e** EBC-SiC/SiC turbine blisk after $N = 994$ tests at $n = 60{,}000$ r/min; **f** Surface morphology of SiC/SiC turbine blisk after test **g** Surface morphology of oxidation products under SEM; **h** Surface morphology of oxidation products under SEM; **i** Damage in the blade; **j** CT section; and **k** CT section

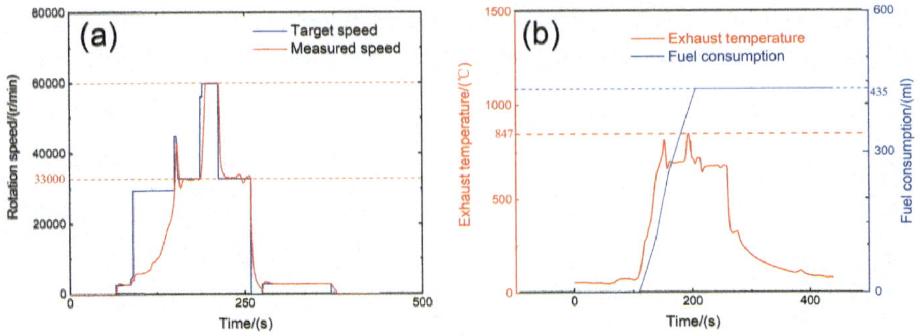

Fig. 5.28 Engine bench test assessment **a** Turbine blisk rotation curves; and **b** Exhaust temperature and fuel consumption curves

During the engine bench test, the partial sections of the blades dropped, as shown in Fig. 5.27i, but did not result in catastrophic failure of the SiC/SiC turbine blisk, indicating that the SiC/SiC turbine blisk has excellent damage tolerance. The NDT inspection of the SiC/SiC turbine blisk was conducted using 3D X-ray CT (XCT) technology. Figure 5.27j and k show typical XCT scans of the SiC/SiC turbine blisk, in which four blades in the horizontal and vertical directions are shown with antioxidant coating sprayed on the surface. The difference between the coating density and the density of the CMC material makes the contours of these four blades significantly shiny. The SiC/SiC turbine blisk adopts the SWS weaving process, and the weaving direction of fiber bundles is mainly along the radial and circumferential directions of the disk, but it does not completely coincide with the radial and circumferential directions, and there are certain angles and dispersions; the overall weaving structure of fiber bundles is more regular, but locally due to the extrusion and deformation of fiber bundles, the local fiber bundles are woven in an irregular manner; there are more unfilled areas of the matrix in the cross-weaving area of fiber bundles and between different fiber bundle layered layers, and there are many pores. In addition, due to the turbine blisk adopts first woven and then the processed manufacturing method, there are more discontinuous fiber bundles in the blade due to the machining, affecting the blade centrifugal force, aerodynamic bending moment load transfer route, the blade strength performance, stiffness performance, and failure damage mode.

5.10 Flight Testing of CMC Turbine Blisk

On January 1, 2022, the SiC/SiC turbine blisk successfully completed its first flight test verification in Zhuzhou, which was also the first aerial flight test of a CMC rotor with an installed platform in China, as shown in Fig. 5.29. This flight verification test is a milestone achievement achieved through the joint efforts of multi-disciplinary directions such as SiC/SiC turbine rotor structure design, strength analysis, manufacturing process, coating preparation, fly-by control and complete aircraft

Fig. 5.29 Flight test of SiC/SiC turbine blisk

test. The research team overcame a number of technical difficulties in SiC/SiC turbine blisk strength design, process preparation, coating preparation, ground test and flight control, and completed the rupture speed test, low cycle fatigue test and ground bench test of the SiC/SiC turbine blisk, which eliminated the potential risk of the flight test and ensured the successful completion of the first flight test. The successful completion of the flight test fully verified the feasibility of the SiC/SiC turbine rotor for gas turbine engine application and accumulated valuable experience for the subsequent development of CMC hot-section components.

5.11 Summary and Conclusions

In this chapter, the design, fabrication and testing of CMC turbine blisk were investigated. The service performance analysis of the turbine blisk was at first analyzed. The spider web structure (SWS) fiber preform was adopted as the reinforcement elements of the turbine blisk. The BN interphase and SiC matrix were deposited on the SWS fiber preform, and during the fabrication, the on-line machining were conducted to control the shape and dimension of the blisk. The environmental barrier coatings (EBCs) were prepared on the surface of SiC/SiC turbine blisk using the atmospheric plasma spraying (APS) method. The XCT scan was conducted to characterize the internal defects in the SiC/SiC turbine blisk. The mechanical performance evaluation, over rotation test and engine bench test were performed, respectively.

- A spider web structure was proposed, and the simulation analysis and test verification showed that this structure can significantly improve the load-bearing capacity and damage tolerance capacity of the CMC turbine blisk.

- A spiral continuous wrap-around disk fabric weaving technique with uniform yarn addition from inside to outside was explored, and the spider web structure was woven to meet the design requirements and low damage requirements.
- The process for the preparation of large-thickness, high-density CMC load-bearing structures was developed, which can significantly improve the density and mechanical properties of the CMC turbine blisk.
- An innovative radial wedge-shaped slip connection structure was proposed to solve the problem of thermal mismatch between the ceramic disk and the metal shaft at high temperature caused by the difference of thermal expansion coefficient.
- The SiC/SiC turbine blisk was installed on an aircraft and completed the first flight testing, which verified the feasibility of the SiC/SiC turbine rotor for gas turbine engine application and accumulated valuable experience for the subsequent development of CMC hot-section components.

References

Bhat BN (2018) Aerospace materials and applications. American Institute of Aeronautics and Astronautics

Cranford SW, Tarakanova A, Pugno NM et al (2012) Nonlinear material behaviour of spider silk yields robust webs. Nature 482(7383):72–76

Freeman TJ, Engel TZ (2019) Turbine blisk including ceramic matrix composite blades and methods of manufacture. 10280768B2, US

Li L (2020) Interfaces of ceramic-matrix composites. Design, characterization and damage effects. Wiley-VCH, Weinheim, Germany

Liu X, Xu Y, Guo X, Luo X, Li L, Zeng Y, Liu C (2022) Fabrication and testing of spider web structure SiC/SiC turbine blisk. J Chin Ceram Soc 50(8):2205–2211

Liu X, Xu Y, Li J, Luo X, Guo X, Hu X, Cao X, Li L, Liu C, Dong N, Liu Y (2023) Design, fabrication and testing of ceramic-matrix composite turbine blisk. Acta Materiae Compositae Sinica 40:1702–1712

Nozhnitsky YA (2018) The problem of ensuring reliability of gas turbine engines. IOP Conf Ser: Mater Sci Eng 302:012082

Ogawa A, Hashimoto R, Zhou F (2001) Rotational test of C/C blisk rotors. In: The 13th international conference on composite materials. Beijing, China.

GPSR Compliance

The European Union's (EU) General Product Safety Regulation (GPSR) is a set of rules that requires consumer products to be safe and our obligations to ensure this.

If you have any concerns about our products, you can contact us on ProductSafety@springernature.com

In case Publisher is established outside the EU, the EU authorized representative is:

Springer Nature Customer Service Center GmbH
Europaplatz 3
69115 Heidelberg, Germany

The manufacturer's authorised representative in the EU is Springer Nature Customer Service Centre GmbH, Europaplatz 3, 69115 Heidelberg, Germany. If you have any concerns regarding our products, please contact ProductSafety@springernature.com

Printed and bound by CPI Group (UK) Ltd, Croydon, CR0 4YY
30/09/2026
02225265-0003